Survival 101 Food Storage:

A Step by Step Beginners Guide on Preserving Food and What to Stockpile While Under Quarantine in 2021

Rory Anderson

Table of Contents

Introduction

In 2020 we faced a global pandemic, mass unemployment levels, and destruction in our supply chains. We have endured a number of different challenges already. Unfortunately, it seems like those challenges are going to continue for the foreseeable future. In the past, similar pandemics had lasted between 18-24 months before everything settled, and it often took years after for things to return to normal. If our current state of affairs turns out anything like historical events have in the past, we will have quite a few months, or even years, left to endure everything that has fallen upon us.

One of the biggest things people are worried about, *you* are likely concerned with, is the destruction in our supply chains. Suddenly, there is a considerable level of uncertainty in what will be available and when. Further, people are becoming increasingly more skeptical about where their products come from and are hesitant to trust products that are being imported from overseas suppliers. What does this mean for you? It means you need to take your supply chains into your own hands.

Modern life is convenient, and it has pampered us, that's for sure. Being able to go to the grocery store and get anything you want, whenever you want, and feeling confident that it will never run out is a luxury that many do not have. It is a luxury that those before us didn't have, either. The reality is, our modern way of life may be lavish, but it is certainly not sustainable. As we are currently seeing, just one thing can completely derail the system and leave hundreds of thousands of people without everyday comforts.

When the system falls apart, what are you left to do?

You either sustain yourself, or you die.

It may sound intense, but it's true. If you do not have food, water, shelter, or any other means necessary to live, you will not live. Money may have bought your way to survival in the past, but if the supply chain is damaged and the products are unavailable to purchase, money isn't going to get you very far. Not to mention, the mass unemployment rates mean that most people cannot even afford to sustain themselves as it is.

The solution to the destruction in the supply chains and the mass unemployment rates is to do it yourself. You need to figure out how you can meet your own needs, and continue meeting them for the long haul. This way, you no longer have to worry about a finicky system that seems to be sitting on the brink of ruin. That starts with knowing how to manage your food supply.

Survival 101: Food Storage will describe, in detail, how you can save money and sustain yourself and your family for the foreseeable future through the power of food preservation. You will discover how you can buy inexpensive produce and meat cuts, preserve them safely, and use them to create meals for months and even years to come. In doing this, you will drastically cut back on your food bill while also ensuring that everyone in your family has plenty of food to eat.

Food storage and preservation may be intimidating to some, especially when you consider the many bacteria and parasites that can live in our food sources. Fortunately, with proper storage and preservation methods, you can destroy any chance of your food containing such things, and confidently consume the food you've stored for yourself and your family. Chances are, you will also discover that homemade preserves are far tastier than those that you can buy at the store. You may even, long after all of this is over, continue with this self-sufficient way of eating. I'll let you decide that for yourself when the time comes.

In the meantime, let me show you how you can confidently stock your pantry, fridge, and freezer with enough food to sustain your family through quarantine and beyond.

Chapter 1

Preservation Methods

Preserving food is done in a variety of ways. Think about every time you have walked through the grocery store and saw food on the shelves. Every single item in that store has been preserved in a way that made it shelf-stable for as long as it needed to be in order for it to be sold and then stored in your kitchen until you were ready to eat it. Yes, each item had an expiry date. All food, no matter how well-preserved, will. However, those expiry dates were *lengthened* by the preservation methods used. Some preservation methods result in shorter expiry dates because the supplier knows that those foods will be consumed quickly. Therefore, lengthy preservation practices are unnecessary. Other methods result in food being able to be stored for months and even years at a time, making them excellent options for long term storage. That is precisely what you want to achieve while in quarantine as it allows you to store food for as long as possible, effectively moving your reliance away from finicky supply chains and into your very own pantry.

How food is preserved depends on what type of food is being preserved and how long you need to preserve it. It also depends on what you have available to you. Not all equipment that is meant for preserving comes cheap or easy, which means that some options may not be available. With that being said, every option *can* be done in your own home; it's just a matter of how much space you have and how much you are willing to invest in the preservation process to get there.

What Types of Preservation Methods Are There?

There are many different types of preservation methods out there! There are also many different types of storage methods. Note that preservation and storage methods are not always the same. For example, when you are canning food, your preservation method and storage method are pretty much one and the same. When you dehydrate food, dehydrating will prepare it for preservation; however, you will need to use an alternate method to store it properly. Vacuum sealed packages are most common for dehydrated foods, especially meats. Sterilized and tightly sealed jars may be used for other types of dehydrated foods, such as dehydrated herbs or vegetable bits.

The most common preservation methods that people use include, water bath canning, pressure canning, dehydrating, freezing, brining and salting, sugaring, smoking, pickling and fermenting, and using ash, oil, or honey to preserve things. Each of these preservation methods will ensure that the food you are preserving is clean, sterile, and shelf-stable so that you can safely store it in your pantry or kitchen for an extended time.

Once you have preserved something, you will need to know how to prepare it after properly. Some things can be served straight out of storage, while others will need to be prepared first. For example, jam or jelly can be eaten right out of the jar, however canned potatoes will still need to be cooked, or canned soup will need to be reheated. Different types of preservation will require different levels of preparation, too. This will especially depend on what has been preserved and at what stage. For example, frozen meat will still need to be defrosted and cooked before consuming because the average freezer does not get cold enough to safely kill off any bacteria and parasites that may be living in raw meat.

How Do You Choose Which Preservation Method to Use?

Choosing which preservation method to use will depend on a few things: what food you want to preserve, what you want to do with the preserved food, when you want to use it, and what tools you have on hand.

Not all preservation methods are made equally, which means not all preservation methods are suitable for all types of foods. A great example would explain the two different types of canning: water bath canning and pressure canning. While water bath canning is excellent for more acidic foods such as jams, jellies, tomato sauces, and certain fruit juices, it is not ideal for less acidic foods such as vegetables, soups, or meats. This is because a water bath canner simply does not get hot enough to safely sterilize the jar and the contents of the jar during the canning process. Pressure canning is used for less acidic foods such as vegetables, soups, and meats because the pressure canner is able to get extremely hot. That heat penetrates the jars and ensures that everything inside of them is sterilized properly.

Aside from which type of food you are planning on preserving, you need to consider what it is that you want to do with that preserve. Most foods will be able to be preserved in a few different ways, so you need to decide what your goal is with the final product. Meat, for example, should be preserved based on how you want to prepare the meat afterward. If you are planning on cooking and incorporating it into a recipe, smoking your meat may not be ideal, unless your recipe specifically requires you to have smoked meat. Consider what the preservation process will do to the ingredient you are preserving and preserve your food based on what you plan on making with that food item in the long run.

Knowing when you want to use preserved food matters, too. Most preservation methods will vary in how long they can safely be consumed for. For example, berries can be frozen for about one year before they begin to lose quality and stop tasting as good. Alternatively, canned fruit can last up to 2 years. Consider storing one type of food in a couple of different ways so it can be used in different types of recipes and for differing lengths of time. This way, you have plenty to last you for a long time, and you can enjoy a varied diet.

Lastly, you need to consider what you have on hand. Especially in the middle of a pandemic where supply chains are damaged, it may be challenging to get exactly what you need in order to preserve food in specific ways. Before you commit to preserving one type of food in any particular way, consider which preservation tools you can get your hands on and then go from there. If you can easily access and afford all of the tools for any chosen method of preservation, then you are good to go. Otherwise, you will need to choose an alternative method that is more accessible and affordable for you.

Are All Preservation Methods Easy Enough for Beginners to Try?

Not all preservation methods are easy, though a beginner can certainly try any method he or she wishes to try. With that being said, it is important that you do not try a recipe that you are not confident in, as you could find yourself making mistakes based on a lack of understanding. These mistakes can lead to the development of harmful bacteria or parasites, which, in turn, can spoil your food and cause serious illness in anyone who consumes the foods.

Always make sure you are confident in the method you are using and that you clearly understand it so that you can safely prepare food for your family. If you are ever uncertain, see if you can recruit the help of someone who knows what they are doing. Watch some YouTube videos which can provide you with more in-depth guidance on each method, or choose an alternative method for preserving your food. As you go, you will find it easier for you to preserve your food, which will help with building your confidence and will support you in adding even more to your home prepared food storage.

Things That Will Destroy Your Preserves

Before you jump into preserving food for yourself and your family, you need to know which things are going to destroy anything you have preserved. Knowing this in advance ensures that you can avoid or minimize these exposures so that you can keep your preserved food for as long as is possible.

The top six things that will destroy your preserves include light, oxygen, moisture, temperature, pests, and time.

Light can create two unwanted situations which are both responsible for damaging your preserved foods. UV radiation has been known to damage certain components of certain foods, causing those foods to break down prematurely, which results in them losing their flavor, texture, and nutritional value. While these foods might fill your stomach, they will not be nearly as enjoyable to eat, and they will not offer as many nutrients as they could have otherwise. Electric lights can still reduce shelf life by increasing the heat that surrounds the food, which can cause it to deteriorate, too. The other element of light that can damage your food is the fact that sunlight can increase oxidization. As the food inside heats up, it begins to produce more gases on its own, which results in more oxygen reaching the food, effectively minimizing its shelf life and possibly spoiling the food altogether.

Oxygen reduces the quality of preserves, possibly to the point of completely spoiling them, in a process known as oxidization. Oxidative spoilage destroys foods that have fat content in them because the lipids oxidize, which results in short-chain carbon compounds being formed. These compounds have a strong odor and flavor and can encourage the development of unwanted bacteria in the food itself.

Moisture reduces the quality of preserves as it can develop in the creation of mold and other spoilages that destroy the food. If moisture gets into food, all sorts of bacteria can develop on or in that food, rendering it dangerous to eat. If you were to eat food that was damaged by moisture in storage, you could fall extremely ill from the bacteria that grew as a result of the moisture.

Temperature reduces the quality of foods when the temperature is too hot or too cold. Too hot can result in the formation of "sweating" inside of the product itself, which can create moisture and effectively cause the development of bacteria. Too cold can cause the item to get freezer burn and can destroy the flavor and quality of the food itself. The ideal temperature for storing a preserved food item depends on what method you have used. However, all foods will not tolerate being kept too hot or too cold.

Pests can completely destroy the quality of preserved food. When a pest has gotten into your food, it is best to throw the food out entirely as pests can bring with them many different risks when they have contaminated your food. Pests can bring about bacteria and viruses by way of feces, urine, and saliva, and they can infest your food with parasites. You should never consume any food item that has been affected by pests as you will be running high risks. Once bacteria or parasites have been introduced to your food, they can multiply rapidly. You do not want to find yourself consuming food that has become contaminated in such a way as the results can be massive, possibly even fatal.

Lastly, time will reduce the quality of your food preserves. Most preserved food will last for many years, possibly even a few decades. However, as time goes on, you will find that those foods start to deteriorate in quality. Eventually, enough time will pass, and the food will no longer be consumable simply because of how much time has gone by. While proper preservation can prevent food from breaking down, eventually, everything breaks down.

Chapter 2

Top Foods to Stockpile

Knowing what food to preserve is essential. There are a few things that will factor into what types of food you should be preserving for yourself and your family. It is important to have a plan for what you will preserve, how much of it you will preserve, and how you will preserve it clearly defined in advance. Going into the process of preserving food without a clear plan can result in you not having enough, having too much of one thing but not enough of another, or having more than you actually need. Not having enough or not having enough of a variety can lead to a poor quality diet, which can ultimately deteriorate your health and put you at greater risk of falling ill. Having more than you actually need can waste your money, which is never a good idea. While it would be better to end up with more than you need, it would be unfortunate to waste a precious resource such as money when this resource is already stretched very thin for most people.

The Master List of Foods to Preserve, Including Superfoods

This list contains all of the foods you should look at preserving for your family, including superfoods that are known for having an abundance of nutrients in them. In an ideal world, you would be able to preserve every single item off this list. However, that may not be reasonable for you as some items may not be easily accessible or affordable, or they may be things that your family simply won't like. We will discuss how you can decide which foods you will preserve for your family next. In the meantime, look through the following lists below to get an idea of what foods you can and should preserve from home.

Meat and Dairy

- White Tuna
- Salmon
- Oyster
- Mussels
- Lobster
- Crab
- Clams
- Beef, Venison, Elk
- Pork, Bear
- Goats, Sheep
- Rabbit
- Pilchards
- Bone Broth
- Eggs
- Milk
- Butter

Vegetables

- Herbs (cilantro, dill, marjoram, pepper, etc.)
- Tomatoes
- Sweet Potatoes
- Carrots

- Green Beans

Fruits

- Mango

- Apples

- Pomegranate

- Berries (strawberries, raspberries, blackberries, etc.)

- Bananas

Miscellaneous

- Hazelnuts

- Dark Chocolate

- Extra Virgin Olive Oil

- Kefir

- Miso

- Raw Honey

- Whole Grain Bread

- Peanut Butter

Pantry Supplies

- Dried Pasta (macaroni, spaghetti, fettuccini, etc.)

- Rice (brown rice has more nutrients, white rice lasts longer)

- Powdered Dairy Products

- Oats

- Sugar (white, brown, syrup, molasses, honey)

- Salts

- Leavening Agents

- Potato Flakes

- Water

Superfoods

- Broccoli

- Garlic

- Kale

- Licorice

- Citrus Fruits

- Bell Peppers

- Ginger

- Spinach

- Yogurt

- Almonds

- Turmeric

- Green Tea

- Papaya

- Poultry

- Kiwi

- Watermelon

- Blueberries

- Acai Berries

- Elderberries

- Sunflower Seeds

- Button Mushrooms

- Medicinal Mushrooms

- Astralagus

- Pelargonium Sidoides

- Spirulina

- Moringa

- Oysters

- Shellfish

Saving Costs Through Preserving Foods

Preserving foods may seem expensive if you look at the up-front cost. Still, the reality is that if you break that cost down over time, you will discover that the overall price is much lower than your average grocery bill. Further, there are many things you can do to help you minimize the costs associated with preserving food.

First and foremost, you need to realize that the "available all the time" mentality is something that was instilled in us by the very supply chains that are presently having a hard time managing our current global affairs. Commercial greenhouses, unnatural growing situations, and access to a worldwide market all resulted in suppliers being able to offer you any food you want, whenever you want, and often at a fairly reasonable price. This is not how it "should" be, though, and when it comes to preserving your own food, following this system can be costly. It can also destroy the flavor and quality of your preserved foods by resulting in you preserving food that is low quality, to begin with, or that may be filled with harsh chemicals that were used to make it grow.

Learning how to buy with the seasons means that you can buy your food as it is in stock from local suppliers. If you can preserve that food as soon as possible, there is a very short window from the item being harvested to the item being preserved. Doing this ensures that your food is the highest quality, is as fresh as possible upon being preserved, and that it lasts much longer. Your food is going to taste better and cost far less, as well.

In addition to buying your foods in season, buy them from local growers and farmers. Local growers and farmers will often offer you the best prices on supplies, especially if you are buying them in bulk, because you are sending so much business their way. Plus, you will be stimulating your local economy. You will know exactly where your food has come from and how it was grown, which are two significant benefits when it comes to surviving a global crisis such as the one we are in right now.

Other ways to save money on sourcing your food for preservation comes from knowing where you can generally go to buy produce and groceries for cheaper. These places are well-known for offering great deals and discounts, all of which can help you save more money while preserving food for yourself and your family.

The following list of resources is sourced from "A Year Without The Grocery Store: A Step by Step Guide to Acquiring, Organizing, and Cooking Food Storage" by Karen Morris. This shares all of the best places to buy your supplies, with the cheapest possible price tag on them:

- Online co-ops
- Bulk food stores and the bulk section of regular stores
- Online food storage companies
- Latter-Day Saints canneries
- Restaurant supply stores

You can also cheapen your items by couponing and by buying in bulk as much as possible, as this often creates a situation where you are paying less per item.

Chapter 3

Effective Stockpiling

How to maximize nutrients, focus on staples with long shelf life, meal prep and preserving leftovers, examples of best foods to stockpile, a mention of what to do if you suffer food allergies or specific health issues relating to food (i.e., Chron's disease, diabetes, etc.)

Determining What Types of Food Your Family Should Preserve

Before you choose what types of food you should preserve, you need to consider what the needs of your family are. Consider your nutritional needs, your nutritional preferences, and any food allergies or intolerances anyone in the family might have. Then, consider the amount of money that you can reasonably budget toward building up your food preserves to ensure that you are getting the best nutritional value that is right for your family, and that fits within your budget.

Start by going through the list above and highlighting every food your family can and will eat. This way, you narrow the list down based on what is reasonable for you to preserve. Next, you are going to consider nutritional value. You are going to want to have at least 3-4 different types of protein stored, 8-10 different types of vegetables stored, 5-7 different types of fruits stored, and a pantry stocked with everything you will need to turn all of your foods into proper meals and get you through the foreseeable future. Highlight several items from each category and favor items that are known for having higher levels of nutritional value. This will give you well-rounded food storage available for you at all times.

Determining How Much Food You Need to Preserve for Your Family

Lastly, you need to consider how much food you need to preserve for your family. This part can be challenging as you are going to need to do some calculations to identify how much food your family needs to sustain themselves for a year.

The absolute minimum amount of recommended food to be preserved per person per year is:

- 365 pounds of vegetables
- 321 pounds of fruit
- 300 pounds of grain
- 182 pounds of meat
- 75 pounds of dairy
- 75 pounds of legumes
- 65 pounds of sweetener
- 5 pounds of salt
- 4 pounds of shortening
- 2 gallons of oil
- 2.5 pounds of leavening agents (yeast, baking soda, baking powder, etc.)

This may sound like a lot, but keep in mind that you do not *have* to store enough food for up to a year. You can store enough food for 3 to 6 months to start and go from there. You can always add more to your inventory every few months to keep you going this way for the long haul. Not everything needs to be added at once, either, nor should it be. Starting with shorter periods may be ideal because it will allow you to get used to storing and using your food preservation stores.

Chapter 4

Planning and Acquiring Your Foods

Properly planning and acquiring your foods ensures that from the very start of your food preservation process, you are organized and ready to preserve everything properly. A proper plan in place ensures that you know what you need to do every step of the way, so that you can acquire all of the right ingredients and supplies for preserving, and that you preserve your food quickly. Once you have obtained your food, you want to preserve it as soon as possible so that it is fresh during the preservation process. This ensures that you do not end up with wasted food that went bad before you were able to get to it. As well, having a clearly organized plan ensures that you do not skip or miss any steps due to being overwhelmed, which means the preservation process will be done properly, and you will not be at risk of getting sick.

Creating a List of All of Your Foods and Supplies

You want to start your planning process by first writing a master list of all of the foods and supplies you are going to need to acquire your foods and preserve them properly. Starting with the food is easiest. Then, research the preservation methods you are going to use and compile a list of all of the tools you are going to need to preserve your food properly. If this is your first time preserving, your food list may be rather large and expensive because you will need to buy things that are not necessarily consumables, such as a pot or a pressure cooker for canning, a smoker, or a vacuum sealer.

Be sure to put every single thing on the list so that you know what it is that you are going to need in order to preserve your food properly. This is going to help you plan your budget and schedule around how you are going to preserve all of your food storage properly.

Once you have added every single thing to your list, estimate roughly how much each item is going to cost you. Knowing how much things will cost will help drastically when it comes to budgeting and planning your schedule. It will allow you to do things based on when you can reasonably afford to do them.

Planning What Will Be Preserved and When

Next, you need to plan what will be preserved, and when. Again, you want to preserve things when they are in season as they will be cheaper *and* fresher, which means your preserves will be higher quality and they should last longer.

The best way to plan out your preserved foods is by what will be preserved and when you need to schedule yourself on a month to month basis for the next 12 months. Write the schedule out based on what types of food will be in season and available for purchase during those months. This will allow you to identify how you can follow the natural harvesting seasons.

Here is a sample of what a preserving season looks like:

- January – April: Winter squash, garlic, peas, beans, carrots grown indoors
- May: Rhubarb, eggs, herbs, winter squash
- June: Garlic, strawberries, snap peas, shelling peas, herbs, onions, fish
- July: Berries, beets, root vegetables, summer squash, beans, fish
- August: Tomatoes, peppers, watermelon, corn, fish
- September: Rhubarb, eggs, herbs, garlic, strawberries, snap peas, shelling peas, herbs, onions, berries, beets, root vegetables, summer squash, beans, tomatoes, pepper, watermelon, corn (most crops are in second harvest by this time)
- October: Apples, pears, quinces, meat

- November: Cabbage, carrots, bok choy, spinach, potatoes, onions, meat, nuts

- December: Organize the root cellar and plan for the next year

Budgeting for Your Preservation Plan

Now that you know what needs to be preserved and when you need to plan your budget. This is going to give you the "final say" in how your preservation season will go since you will need to be sure that you can reasonably afford to preserve everything you have chosen.

Because every month is going to have such different harvests and preservation needs, you are going to want to know how much each month will cost you. Consider both the foods you will need to buy *and* the resources you will need to make that happen. This way, you know an exact calculation of what that months' worth of preservations would cost you.

Once you know the cost of preserving, you need to factor it into your annual budget. If you have not already made one, now is the time to start. See if the cost for preserving as much as you need and want is feasible and, if it isn't, see where you can trim costs. You can rely on the resources mentioned in *Chapter 2: Top Foods To Stockpile* as a way to cut costs. You can also go to local farms and farmer's markets, where direct suppliers will often offer great discounts on items that are ready to be sold. In some cases, you may be able to find some items that are nearing the end of their shelf-life in fresh form for cheaper because they are nearly ready to be tossed. These types of foods are often good for canning because canning will preserve them for much longer.

Another way you can adjust your canning season to fit your budget is to swap out more expensive items for cheaper ones. For example, lobster and crab can be quite expensive, so you might swap them out for salmon and tuna instead. Certain berries like elderberries also tend to be more expensive, so you might consider swapping them out for blueberries. Making these types of swaps in your preservation plan creates a cheaper preservation process, effectively helping you save as much money as possible.

Don't just look at saving money on the food itself, either. Often you can find resources on sale or even available at lower prices second hand. Just make sure that anything you buy, whether it is new or used, functions properly, and can create the results you desire.

Acquiring All of the Resources You Will Need First

Before you begin your preservation process, you are going to want to acquire your resources first. Look into what types of preservation methods you will be doing in a given month and stock up on everything you will need for those methods. If you are canning, make sure you have a proper stockpot, water bath canner, jar holder, jars, pectin, and anything else you need for canning. If you are going to be smoking your food, make sure your smoker has fuel, you have the proper wood chips, and you have a vacuum sealer with fresh bags to use to help you store your smoked meats after.

Before you go to purchase your food supplies, make sure you double-check all of your resources. You want to be sure that you have everything you need *and* that everything you have is in proper working order. This way, you do not find yourself without something partway through making a recipe and are forced to either stretch yourself thin to get it or screw up your entire preservation recipe because you do not have it. Remember, with most preservation methods, you will not have the option to substitute various things because doing so can mess up the preservation process and introduce harmful bacteria into your foods. You must have exactly what you need and in perfect working order before you bring your food home so that you can safely preserve your harvest every single time.

The other resource you need is one that will likely not cost anything unless you choose to invest in a cookbook. That is, you are going to need recipes. Recipes can be found online or in cookbooks that are designed for your preferred method of preservation. Be sure to rely on tried and tested recipes and not just any recipe, as the recipes that are truly intended for preservation will be written with safety considerations in mind. Do not purchase any food supplies until you have compiled all of the recipes you will be making, as this will ensure that you are able to create a complete grocery list beforehand and that you can go out and get these items all at once. That way, you are not left running back and forth to the store, trying to pick up everything you forgot about the last time you went.

Bringing Home Your Food Supplies

Once you are sure you have all of your preservation supplies and resources available, you are going to need to bring home your food supplies. Bringing home your food supplies should be done in a three-part process.

First, you are going to want to locate where you are going to store your food supplies for the short term as you process everything. It may take a few days for you to process everything completely, so you need to make sure you have room on your shelf, in your fridge, or in your freezer to hold onto the excess supplies until you are able to get to it. If you do not know where you will be storing your food yet, do not bring it home right away as it could lead to you destroying the quality of the food before you can safely get it into your preservation system.

Second, you are going to want to source where you are getting your food from. At this point, especially if you're going to be as frugal as possible, you are going to need to shop around to make sure you get the best deal. Look around to see what is available and find the best deals possible; if you can, factor in the cost of sale pricing, coupons, and other deals that may help you save more money. If you are going to be buying more than one thing in a month, make sure you consider the convenience factor of picking everything up. You don't want to drive around to several different stores and locations with your food sitting in your vehicle, especially in warm months. This will waste time, gas, and possibly cause your food quality to deteriorate while it sits in the car. Try to get everything from one, two, or maybe three different locations at most.

Third, you are going to have to get your food home. For the most part, you should only need to stack everything in your vehicle. If, however, you do not have a vehicle or if you have more than you can reasonably put in your vehicle, you are going to have to find an alternative way to bring your produce home. When pandemic orders are not in effect, you may be able to rely on an Uber or taxi to help you. When pandemic orders are in effect, it would be safer to rely on a family member or a friend or to have the store deliver your food to your house if possible. However, that will cost more as you will have to pay a delivery fee for your food.

Creating Your Recipe Plan

If you are going to be processing all of your food over a day or two, you can create your recipe plan once your food is at home. If you are going to be processing food every week for a month, you are going to want to do this ahead of time so that you only buy the food that you will be able to process in the next couple of days after bringing it home.

Your recipe plan essentially states which recipes you are going to use and when you are going to use them in the day. Having a detailed recipe plan ensures that you stay organized on processing day, so you do not lose track of what needs to be done or what has already been completed. I suggest organizing your recipes based on the major ingredients in them, or the types of recipes you are doing so that you can keep it clean and organized. For example, do all of your recipes with tomatoes in a single day or over two days; that way, you are able to process all of your tomatoes and have them done finished. If you are going to be using two major ingredients in a single day, work with one first, and then the next.

Keeping your recipes organized means you do not have to keep putting things away and taking them out again to do your next recipe, and then your next one, and so forth. This way, you save time and can move from one recipe to the next quickly, and then you can clean up all at once. Or, on a particularly large day, you may want to plan for cleaning periods in between your recipes so that you can bring the organization back to your kitchen. Remember, disorganized kitchens can be a hazard for cuts, burns, and other injuries, so you want to keep your kitchen as clean and calm as possible.

Executing Your Food Preservation Plan

Lastly, you need to execute your food preservation plan! Once you have made one, it is important that you do not stray away from it and that you stay on track with the plan you made for yourself. Only make what you put on the plan, and in the exact order you planned. This way, you do not end up running out of ingredients or finding yourself confused about where you are in the plan and missing important recipes.

The only time you should alter your plan is if you run into trouble that seriously stops you from being able to proceed with a certain recipe or ingredient. In this case, you should stop immediately (or as soon as you can if you are mid-process) and decide on a new plan that factors in safety and efficiency. Otherwise, everything should be done per plan. If you do run out of time in one day, simply carry on the next day.

Chapter 5

Storing Different Types of Food

As I mentioned previously, every single type of food has its own methods for being stored. Storing food properly for the type of food it is, is essential because this is how you are able to prevent that food from spoiling. It is imperative that you never attempt to preserve or store food in a method that is not approved for that variety of food because you are running the risk of allowing harmful bacteria, such as *Clostridium Botulinum,* or the bacteria that causes Botulism.

Below, you will discover the best methods for storing each different type of food safely. Again, you are going to want to pick which kind of preservation you use based on what you plan on using that ingredient for later. For example, if you're going to be able to cook a steak, later on, turning all of your beef into beef jerky is not the right choice. Consider how you intend to use your harvest before preserving it so that you can use it for its intended purposes later on.

Meat

Meat can be preserved in a variety of ways, and how you preserve it will affect the taste, texture, and how it can be used later on.

The following ways are approved for storing different types of meats:

- Smoking:
  - This method is most common in red meats, though it can be done in poultry, too. It will create a drier, chewier texture in the meat itself. You will also notice the flavor of the wood that was used to smoke it. Smoked meat should be consumed sparingly as the smoke itself contains carcinogens.

- Salting:
  - Salting, or curing, is an old school preservation method that can still be used. Salt requires more time and effort, but it is cheap. It is often used for meats like bacon or pastrami. You can also salt then smoke meats as a way to enhance flavor.
- Brining:
  - Brining is another traditional preservation method. Brine is made of water, sugar, and salt, and it is poured over meat as the meat is held down at the bottom of the brine using a weighted crock. This method can also take up a lot of space.
- Canning:
  - You cannot water bath can meat, but you can pressure can it. Meat is not acidic enough for water bath canners. Any type of meat can be pressure canned, and this method is most simple as it requires no further work to preserve it. When you are ready to consume the meat, you open the jar and reheat the meat itself. Canned meat is cheap to make, and the preservation process will turn even tougher meat cuts into tender, delicious meat.

- Dehydrating:
  - This method is easy, cheap, and healthy. Electric dehydrators or solar dehydrators can be used. You should buy the largest dehydrator you can afford, so you are not running several batches through a smaller dehydrator. Dehydrated meat can be eaten as is. It will need to be stored in vacuumed sealed packages until you are ready to eat.
- Stored In Lard:

- If you are storing an animal with a lot of fat, like a pig, you can store it in a crock with lard. This method works the same as brining, except you are using lard instead of brine. The lard completely surrounds the meat and prevents air from getting to it. It is cheap and effective and requires no special equipment.

- Freeze Drying:
  - Freeze driers are expensive, so this is not a practical method for most people. However, freeze-dried meat can be stored for a long period of time, it is lightweight, and the food retains nearly all of its nutrition. These foods are also great if you need to evacuate in case of an emergency because they are so light. Freeze-dried meat can be rehydrated by adding it to a liquid and cooking it, though the texture will be very different from other meats.

- Freezer:
  - The freezer is a great way to store meat. Make sure you store it in heavy-duty freezer bags or containers so that it does not become freezer burnt. If meat does become freezer burnt, however, it can be used to make a stew or other similar recipes, so long as the freezer burn is not too bad. Frozen meat stays closest to its raw form, which means it can easily be dehydrated and incorporated into virtually any recipe.

Seafood

Seafood needs to be preserved carefully as it is more finicky and more prone to bacteria and parasites that are unsafe for humans. It is important to properly clean and preserve your seafood to avoid contamination or getting anyone sick in the process.

The following methods are approved for safely preserving seafood:

- Freezing:
  - Freezing is generally the most popular method for storing seafood. For lobster, freezing is the only safe way to store seafood. Seafood tends to be especially prone to the development of bacteria, and the meat is so delicate that preserving it in other ways can damage the meat. There are, of course, exceptions to this, however.
- Brining:
  - Seafood can be brined in salt before being stored. Varieties like herring, salmon, rockfish, and mackerel can be brined and kept for up to 9 months using this method.
- Drying:
  - Drying is a popular way of storing fish, and it can be stored for up to 2 months this way. Most varieties of fish can be dried. Once a fish has been dried, it can be crushed as a condiment or rehydrated by adding it to a soup recipe.
- Smoking:
  - Smoking a fish is a great way to preserve it while creating a delicious texture and flavor. Most fish can be smoked, though salmon and oysters are particularly common varieties for smoked seafood options.
- Canning:
  - Fish should only ever be canned in a pressure canner as water baths will not safely can them for you. Fish are not acidic enough for water bath canners. Virtually every variety of fish can be canned except for lobster. Canning a fish will make it incredibly juicy and will make it delicious to enjoy later on. Plus, it retains its ability to be used in a variety of recipes this way.

Vegetables

Vegetables are fairly easy to preserve, though different varieties will require different types of preservation methods. Leafy greens, for example, are hard to preserve because they are so full of water content and delicate that they can quickly deteriorate in most storage methods.

The following methods are approved as safe ways to preserve your vegetables:

- Drying:
  - Dehydrating vegetables removes the water content from them so that they can be safely stored. You need to make sure dehydrated vegetables are completely dehydrated so that they do not develop bacteria. Food needs to be at least 95% dehydrated before it is considered shelf-stable, and it will still need to be stored in a proper container.

- Canning:
  - Like meat and seafood, vegetables are not acidic enough to endure water bathing. You will need to use a pressure canner to can your vegetables properly. Most vegetables can be stored this way safely for quite a long period of time. Note that vegetables stored this way will be quite moist, so you will not be able to retain the fresh, crunchy texture this way.

- Pickling:
  - Pickling vegetables requires you to have an acidic base such as vinegar to store your vegetables in. Not all vegetables will taste right when they are pickled, though you can pickle a wide variety of things. Cucumbers, asparagus, bell peppers, beets, carrots, cauliflower, fennel, ginger, green beans, mushrooms, onions, parsnips, hot peppers, radishes, rhubarb, squash, and turnips can all be pickled, among other things.

- Fermenting:
    - Fermenting is a process where you convert carbohydrates to alcohol or organic acids. You do this using salt, whey, or another starter culture to a vegetable as a brine and letting the food sit in the brine so that it ferments. Fermented foods are incredibly healthy, as long as they are done right, and they are quite easy to do. Most foods should be able to be fermented relatively easily.
- Freezing:
    - Freezing is a good way to retain the freshness of your vegetables. When you freeze vegetables, you may need to blanch them or flash freeze them in order to freeze them effectively. In some cases, you may need to do both.
- Oil Packing:
    - Oil packing vegetables means that you are essentially storing vegetables in oil. Oil creates anaerobic conditions, meaning that virtually no air can get to the vegetables inside, which results in them being stored safely. This process can be used on tomatoes, eggplants, herbs, onions, and olives.
- Salting:
    - Like meat, vegetables can be salted to cure them, too. Salting vegetables is done in a different means as salting meat because the content of vegetables is so much different. However, it does cure and preserves them all the same.

Fruit

Fruit is similar to vegetables in that it is fresh and grown out of the ground. However, fruit is more acidic than vegetables are, which means fruits can be preserved in a couple of different ways, too.

The following methods are approved as safe methods for preserving fruit:

- Drying:
    - Like vegetables, fruit can be dehydrated. Dehydrated fruit can be consumed as is, it can be added to recipes, or it can be rehydrated and consumed that way. However, rehydrated fruit will not taste the same as it did before it was dehydrated.

- Canning:
    - Fruit is one of the only things that can be done safely in a water bath canner, aside from some recipes that are meant for pickling. With fruit, canning can be used to preserve the whole fruit, or it can be used to allow you to break fruit down into other things such as jams, jellies, juices, salsas, and sauces, which can then be canned and stored.

- Pickling:
    - Believe it or not, fruits can be pickled! Blueberries, cherries, grapes, mangoes, peaches, strawberries, tomatoes, and watermelon can all be pickled and stored.

- Fermenting:
    - Fruits are less likely to be used in ferments as vegetables are unless you are making wine, of course. Fruit can be fermented, however. It typically takes much less time to ferment fruit because it is already acidic. It can take up to 48 hours to ferment fruit, while it can take several days or even a couple of weeks to ferment vegetables.

- Freezing:
    - Freezing is a great way to store your fruits. Most fruits will freeze well. Fruits will not need to be blanched before being frozen; however, it can benefit to cut them up and flash freeze them first. Certain fruits like citrus fruits, tomatoes, bananas, and watermelon do not freeze well because they become mushy and

gross once they are defrosted. If, however, you only plan on using them in smoothies or in baking recipes, it may not be such a big deal to freeze them. However, there are better means of storing them.

Dairy

Dairy products can be preserved in a variety of ways. With that being said, not all dairy products are the same, so you are going to need to preserve them properly to be able to use them. As well, you are seriously going to need to consider what you want the dairy for, as most preservation methods will alter the flavor and usages drastically.

The following are approved and safe methods for preserving and storing dairy products:

- Freezing:
    - Freezing is the most dependable way to extend the shelf life of dairy products, including eggs. With that being said, once you freeze a dairy product, it will separate, and it will never taste quite the same, even if you stir it up afterward. It will still be good for baking, or for survival means, however.
- Dehydrating:
    - Eggs can be dehydrated in a food dehydrator and, when done this way, can store for months. Cheese can also be dehydrated. Eggs will need to be recooked before they can be used, while cheeses can be crushed into a powder and dusted on top of foods for flavoring means.
- Wood Ash:
    - Wood ash will increase the shelf life of both eggs and cheese. Eggs stored in hardwood ash can last for up to a year, while firm cheeses can be stored for

about three months this way. Note, however, that cheeses will start to taste like the ash after about three months.

- Oil Storage:
  - Softer cheeses like mozzarella and feta cheese can be submerged in olive oil and stored for quite a while, although they will still need to be refrigerated. If you oil eggs first, they can be stored at room temperature for up to 5 weeks.
- Wax Storage:
  - Hard cheeses like cheddar can be coated in wax and stored outside of the fridge in a cool, dry place.

Bulk Foods

- Air Tight Containers:
  - Most bulk foods just need to be stored in sterile, airtight containers. For smaller amounts of an ingredient, you can use sterile mason jars. For larger ones, you can invest in larger containers that are usually used in grocery stores or restaurants. Containers are usually made from HDPE or high-density polyethylene and are easy to sterilize and keep airtight.

Dried Foods

- Air Tight Containers:
  - Like bulk foods, airtight containers are often the best way to store dried foods. Sterile mason jars or larger bulk containers can be used for any dried food.
- Vacuum Sealer:
  - Vacuum sealers are great for dried foods, especially those that may still have a small amount of moisture in them, such as raisins or other dried fruits or

meats. Vacuum sealers are the best way to ensure that absolutely no air gets into your food so that you can keep it stored and safe for a long period of time.

Chapter 6

Canning

Canning is the most popular method of storing, aside from freezing, for the average household. Canning is easy to get into because the materials are quite affordable, they are abundant, and these types of foods can last for an extended period of time. Another benefit of canning is that so many foods can be preserved this way, which means you do not have to invest in several expensive materials to get started. If you are new to preserving food, getting started with canning is a great idea as it allows you to jump right into preserving many different types of foods. Plus, the mason jars you buy for canning can also be used for preserving other foods such as dried ingredients, too.

If you are going to get into canning, you should start with a pressure canner. You can always purchase a water bath canner at a later date, but pressure canners will do far more varieties of food for you. These days, there are electric pressure canners available which tend to be easier to manage if you are new to canning because they automatically control their temperature, you do not have to do it for them. However, these pressure canners are quite a bit smaller, so it will take longer for you to get everything done.

The methods for canning are similar, but there are some key differences between the two styles. Be sure to research the specific method of canning you will use beforehand and educate yourself on how that method works before getting started so that you do not injure yourself or damage your food in the process.

Water Bath Canning

Water bath canning is a popular canning method among hobbyists who like to make jams, jellies, salsas, and other similar recipes. A water bath canner is a large pot that holds at least 7-

quart jars while allowing them to be safely submerged in water with 1-2 inches of water above the heads of the jars. The canners have a rack that is submerged in water that holds the jars, so they are not directly against the bottom of the pot by the heat source. The jars then boil for a certain amount of time, based on what recipe is inside of the jar and what altitude you are at.

What Foods Does Water Bath Canning Work for?

Water bath canning works for most fruits, as well as fruit juices, salsas, tomatoes, pickles and relishes, chutneys, sauces, pie fillings, vinegar, and certain condiments.

What Materials Do You Need for Water Bath Canning?

In order to water bath can you will need:

- Mason jars with new lids (or reusable lids. You cannot reuse non-specified reusable lids as they will not reseal properly.)
- Water bath canning pot
- Jar lifter
- Large stockpot
- Stir stick
- Timer
- Canning funnel
- Pectin

How Does The Water Bath Canning Process Work?

Water bath canning works by first preparing a recipe to place inside of your jars. For jams, jellies, salsas, chutneys, sauces, and so forth, you will need to prepare a specific recipe for your canning method. For other things, such as storing sliced fruit, you will need to make a specific syrup to store the fruit in.

Once you have made your recipe, you will need to place that recipe inside of your sterilized jars, add the lids, and boil them in the water bath canner for a specific period of time that's based on the recipe inside and your altitude. This will kill any *Clostridium Botulinum* bacteria inside of the jars so that everything within them is safe. You will remove the jars afterward using a jar lifter and then place them on your counter. Then, you will leave them on the counter overnight to cool. You will check the lids in the morning to make sure they are all properly sealed. Do not tighten the rings at this point; just leave them as they are. If any lids pop off, you will consume those right away. Others can be stored on the shelf for 1-2 years.

Note that with canning recipes, jars can be smaller than defined in the recipe but not larger. The amount of time that jar spends in the water bath will remain the same unless otherwise noted.

Simple Strawberry Jam

8 x 250mL jars.

What You Need:

- 5 cups Crushed Strawberries
- 7 cups Granulated Sugar
- 4 TBSP Lemon Juice
- ½ TSP Butter
- 1 Pack Pectin

How to Make It:

1. Place eight sealed jars in your water bath canner. Fill with water to cover the canner with 1-2 inches of water. Remove the lids, set the screw bands aside, and drop the sealing discs into the hot water, but not on top of the jars. Simmer at 180F as you prepare your jam. (This will sterilize the jars.)

2. Prepare strawberries and crush them one layer at a time. Measure out 5 cups of crushed berries. Measure and set aside granulated sugar.

3. In a pot, combine strawberries, lemon juice, and butter. Dissolve pectin into the mixture.

4. Bring the mixture to a full rolling boil, then add all of the sugar. Stir constantly while returning the mixture to a full rolling boil that does not disappear as you stir it. Let it boil for 1 minute while stirring. Remove the entire mixture from the heat. If foam forms, skim it off.

5. Remove jars from the water and carefully use a tea towel to dry the jar, being cautious not to burn yourself. Ladle the jam into the jar, leaving ¼" headspace. Use a nonmetallic utensil to remove air bubbles. Adjust headspace if needed by adding more jam. Wipe the rim clean. Remove a sealing disc from the water, dry it, and carefully center it over the jar. Screw the screw band on until it is fingertip tight. Return the jar to the canner. Do this with all of the remaining jam.

6. Ensure the jars are covered with at least 1 inch of water. Cover the canner and bring it to a full rolling boil. Boil the filled jar for 10 minutes for altitudes of up to 1,000 feet.

7. When done, remove the lid, turn off the heat, and wait 5 minutes. Without tilting them, remove the jars from the canner and place them on a layer of two tea towels on the counter, so they don't scald your countertop—cool upright for 24 hours. Do not retighten the screw bands.

8. Check seals after 24 hours. Sealed discs will curve downwards and will not move at all if you press on them. Remove screw bands, wipe and dry them and the jars, and replace the screw bands on the jars loosely. Label and store your jars.

Tomato Salsa

Makes 10 x 250mL jars or 5 x 500mL jars.

What You Need:

- 7 cups Tomatoes, chopped
- 2 cups Onions, chopped
- 1 cup Green Bell Pepper, chopped
- 3 Cloves Garlic, minced
- 8 Jalapenos, chopped
- 156mL Tomato Paste
- 175mL White Vinegar (3/4 cup)
- ½ cup Cilantro, chopped and lightly packed
- ½ TSP Cumin, ground

How to Make It:

1. Place sealed jars in your water bath canner. Fill with water to cover the canner with 1-2 inches of water. Remove the lids, set the screw bands aside, and drop the sealing discs into the hot water, but not on top of the jars. Simmer at 180F as you prepare your jam. (This will sterilize the jars.)

2. Blanch tomatoes. When done, peel them and remove the seeds. Coarsely chop them. Measure 7 cups and set aside.

3. Finely chop and remove seeds from jalapenos. Set aside.

4. In a large saucepan, combine tomatoes, green pepper, onions, jalapenos, garlic, vinegar, tomato paste, and cilantro. Bring to a gentle boil. Stir occasionally—Cook for about 30 minutes, or until salsa reaches desired doneness.

5. Remove jars from the canner and dry them, taking care not to burn yourself. Dry the sealing discs, too. Ladle hot salsa into the jars, leaving ½" headspace. Use a nonmetallic utensil to remove air bubbles and add more salsa if needed. Wipe jar rim, center the sealing disc, and screw on a screw band until it is fingertip tight. Return the jar to the canner.

6. When the canner is full, cover all jars with at least 1" of water. Add the lid and bring the water to a full rolling boil. Boil for 20 minutes for altitudes up to 1000 feet.

7. Without tilting them, remove jars and place them on two-layered tea towels on the counter and leave them for 24 hours. Do not tighten screw bands. Check the seals in 24 hours. If any are not sealed, transfer them to the fridge and use them first. Dry off all jars and loosely replace screw bands.

Sliced Peaches

Makes 4 x 500mL Jars or 8 x 250mL Jars

What You Need:

- 9 Large Peaches, sliced
- 1.5 cups Granulated Sugar
- 8 cups Water

How to Make It:

1. Place four sealed jars in your water bath canner. Fill with water to cover the canner with 1-2 inches of water. Remove the lids, set the screw bands aside, and drop the sealing discs into the hot water, but not on top of the jars. Simmer at 180F as you prepare your jam. (This will sterilize the jars.)

2. Cut a shallow "X" into the bottom of each peach. Boil a large pot of water and fill a large bowl with ice water. Boil the peaches for 3 minutes, then remove them straight to the ice water. When peaches are able to be handled, peel them.

3. Slice the peaches and discard the pit and skin. Divide the sliced peaches between the jars.

4. In a medium pot over medium heat, combine sugar and 8 cups of water. Simmer and stir until sugar has completely dissolved.

5. Remove jars from the water bath canner and seal discs. Wipe them dry without burning yourself. Pour syrup over peaches, leaving ¼" headspace. Gently tap jars to settle the peaches and syrup. Add more syrup if it is necessary to reach the ¼" headspace.

6. Return jars to water bath canner, making sure there's 1" of water over the jars. Bring the water to a boil and let boil for 25 minutes for altitudes up to 1,000 feet.

7. Remove jars from pot and let them sit on two-layered tea towels on the counter for 24 hours. Do not tighten screw bands. After 24 hours, check the seals. Any that failed to seal should be kept in the fridge and eaten first. Dry jars and screw bands and loosely return screw bands to sealed jars and store.

Pressure Canning

Pressure canning is a method of preserving that uses a large pot with a lid that is tightly sealed on top of it. Pressure canners reach far higher temperatures than simply boiling water in a pot.

This is the only way to reach such temperatures with water effectively. These temperatures are required to kill *Clostridium Botulinum* so that the foods inside of the jars are safe. In foods that are not as acidic, such as meats and vegetables, this is necessary as it is more challenging to kill *Clostridium Botulinum* off of these foods than it is off of those that are more acidic, such as fruits.

What Foods Does Pressure Canning Work for?

Vegetables, fruits, meats, poultry, seafood, soups, and broth can all be safely preserved using a pressure canner.

What Materials Do You Need for Pressure Canning?

A pressure canner will need to be set up with the lid, the rack, and about 2-3 inches of water inside of it.

How Does the Pressure Canning Process Work?

Cans are placed on the rack within the canner, and the pot is sealed before it is heated to 240F. It will need to be maintained at that temperature for a period of time, based on what you are canning and what altitude you are at. When that time is reached, you will use the vent to release some of the pressure from inside of the can. Then you will carefully remove the lid, remove the jars using proper jar lifters. Then, you will let the cans sit overnight, and you will check them the next day. Do not tighten any of the lids. Any jars that did not properly seal should be consumed right away, while everything else can be stored for 1-2 years. Be very careful with pressure canners as they get much hotter than boiling water and can cause disastrous injuries. Further, the pressure can blow the lid off and cause serious damage to your house if you are not careful.

Note that with canning recipes, jars can be smaller than defined in the recipe but not larger. The amount of time that jar spends in the pressure canner will remain the same unless otherwise noted.

Canned Ground Meat

As Many As You Want. Pints or Quarts.

What You Need:

- Ground Meat
- Boiling Broth Or Water
- 1 TSP Salt Per Pound Of Ground Meat

How to Make It:

1. Sauté meat until just done. Add 1 TSP of salt per pound of meat you are using. Drain away any excess fat.

2. Fill clean jars with meat. Cover them in boiling broth or boiling water. Leave 1" headspace. Center sealing discs and screw bands until fingertip tight.

3. Process at 11 pounds of pressure in a dial gauge canner for 75 minutes if you are canning pints or 90 minutes if you are canning quarts if you are up to 2,000 feet altitude. Process at 10 pounds of pressure in a weighted gauge canner for 75 minutes if you are canning pints or 90 minutes if you are canning quarts if you are up to 1,000 feet altitude. Adjust pressure and cook time based on your altitude.

Canned Green Beans

As Many As You Want, Pints Or Quarts.

What You Need:

- Beans
- Boiling Water

How to Make It:

1. Wash young beans thoroughly. Chop off the stem, blossom ends, and "strings" if any appear. Leave whole or cut into 1" pieces. Make sure whole beans will fit into jars while being able to leave 1" headspace.

2. Pack beans tightly in clean, hot mason jars, leaving 1" headspace. Cover with boiling water, still leaving 1" headspace.

3. Process at 11 pounds of pressure in a dial gauge canner for 20 minutes for pints and 25 minutes for quarts if processing at altitudes of up to 2,000 feet. Process at 10 pounds of pressure in a weighted gauge canner for 20 minutes for pints and 25 minutes for quarts. Adjust pressure and cook time based on your altitude.

Canned Broth

As Many As You Want, Pints Or Quarts.

What You Need:

- Bone Broth (make it in advance by simmering a whole poultry carcass or fresh trimmed beef bones in a stockpot. Simmer poultry bones for 30-45 minutes, simmer beef bones for 3-4 hours.)

How to Make It:

1. Prepare your broth by simmering a poultry carcass for 30-45 minutes or beef bones for 3-4 hours. If using beef bones, rinse them before placing them in the water to prepare your stock with. After recommended cook time, remove the bones and let your broth cool to room temperature. Skim off any fat and discard—strain broth to remove bits of bone. If there was meat on the bones, you can remove it and add it back to the broth for flavor if you want. Reheat broth to boiling.

2. Fill clean jars with hot broth, leaving 1" headspace. Center sealing discs and screw on screw bands to fingertip tight. Process at 11 pounds of pressure in a dial gauge canner for 20 minutes for pints and 25 minutes for quarts at altitudes up to 2,000 feet. Process at 10 pounds of pressure in a weighted gauge canner for 20 minutes for pints and 25 minutes for quarts at altitudes up to 1,000 feet. Adjust your pressure and times based on your altitude.

Chapter 7

Dehydrating

Dehydrating is a form of food preservation that involves removing the moisture content from foods so that 5% or less moisture remains in the food itself. Food is dehydrated in proper dehydrators, which are designed to wick away moisture and dry the foods inside. While you can use the sun to dehydrate foods, it is not recommended as you cannot control the temperature or the rate at which food is being dehydrated. It can take anywhere from a few hours to an entire day to dehydrate food. Dehydrated food will have different textures based on what was dehydrated, and how thick it was sliced before dehydrating. Bananas, apples, and onions, for example, will dehydrate to the point where they are crispy. Slices of meat, apricots, and mangoes will retain a somewhat chewy texture, even after they have been dehydrated. All dehydrated foods will need to be properly stored in airtight containers afterward to prevent spoilage. Foods that retain closer to the 5% moisture allowance should be kept in vacuum-sealed bags until you are ready to use them to prevent spoilage.

What Foods Does Dehydrating Work for?

Most fruits, vegetables, and meats can be dehydrated. Dairy products like cheeses and eggs can be dehydrated, too. You can dehydrate foods with the intention of eating them dehydrated, of rehydrating them later, or of adding them to baked goods later on. Certain foods, like eggs, will need to be cooked even if they have been dehydrated as they still run the risk of causing illness if you eat them raw.

What Materials Do You Need for Dehydrating?

To dehydrate, you are going to need a handful of materials. These materials will prepare the foods for the dehydrator, allow you to dehydrate the foods, and allow you to store the foods after they have been dehydrated.

They include:

- An electric dehydrator (get the largest one you can afford without compromising on quality)
- Dehydrator tray liners
- Cutting board
- Sharp knife
- Vacuum sealer with bags
- Blender

How Does the Dehydrating Process Work?

The dehydrating process works by first preparing foods to go into the dehydrator. In order to work efficiently, foods need to be sliced into thin slices so that they can dehydrate relatively quickly. If a food is too thick upon going into the dehydrator, you will end up either waiting excessive amounts of time for the food to dehydrate or finding yourself unable to completely dehydrate the foods to less than 5% moisture, which means bacteria will grow. Once your food is sliced, you will place it on the dehydrator trays, turn on your dehydrator, and dehydrate your foods until they have less than 5% moisture remaining. The amount of time required will vary depending on what you are dehydrating, so it is best to start all recipes in the morning to avoid having to check on your food into the later hours of the night and early hours of the morning.

If you are making something like dehydrated eggs, you are going to want to use a dehydrator tray liner, which is used to prevent wet ingredients from dropping through the tray. This way, they dehydrate correctly. Dehydrator tray liners can also be used to make fruit leathers. To make fruit leathers, you will blend up a fruit recipe in the blender until it is completely smooth and then pour it over the dehydrator tray liner before dehydrating it. Once it's done, you will have delicious fruit leathers.

Dehydrating can be done in an oven, too, although it will use up a lot more energy, and the foods may lightly cook in the process. While this does work, it is ultimately best to use a proper dehydrator, which can keep the right temperatures and conditions inside of the appliance for proper, safe dehydrating.

Dehydrated Potato Flakes

1 x 1 Pint Jar

What You Need:

- Five potatoes, peeled and chopped

How To Make It:

1. Add peeled and chopped potatoes to a pot and cover them with water. Boil the potatoes for about 10-15 minutes until they are fork-tender and ready to be mashed.
2. Move soft potatoes into a bowl, taking care not to add any water to the bowl in the process. Mash the potatoes until they are smooth. Refrain from adding anything to flavor the potatoes or soften the potatoes at this point, such as herbs or milk.
3. Lay potatoes on a fruit roll sheet on your dehydrator tray and dehydrate on 145F for 6 hours.

4. Break potato sheet into chunks, then blend them in your blender until they are ground into flakes. If you make your flakes too fine, they will be sticky when you cook them, so refrain from blending them too much.

5. Store potato flakes in a clean, airtight container in a cool, dry spot for up to 6 months.

Dehydrated Mango Slices

About 1-2 pounds of dehydrated mango.

What You Need:

- 5 ripe mangoes, peeled and pits removed
- ¼ cup lemon juice
- 1 TBSP raw honey

How to Make It:

1. Slice mangoes into thin strips.

2. Stir honey and lemon juice together. Dip thin mango strips into the mixture, shake off excess juice, and place them on dehydrator sheets.

3. Dehydrate for 10-12 hours at 135F. Around 8 to 9 hours, begin checking your mangoes for doneness. The exact time it will take depends on how ripe your mangoes were, how humid it is in your home, and how thin you sliced the strips.

Tasty Fruit Roll-Ups

Approximately 8, depending on what length you choose to cut yours.

What You Need:

- 5 cups of fruit, any fruit you prefer (you can also add some vegetables for added nutrition)
- ¼ cup raw honey

How to Make It:

1. Clean your fruits and vegetables for processing. Remove seeds and stems. The peels are nutritious, so you can keep them on for your fruit roll-ups; however, you can remove them if you decide you do not like the taste or texture. For foods where peels are not typically eaten, such as bananas, oranges, or pineapples, remove the peel.

2. Place all of your fresh fruits and vegetables into a blender with honey and blend until it is completely smooth. If you prefer even smoother textures, you can use a juicer to prepare your fruits and vegetables. However, there will be a lot of waste, and much of the pulp contains a great deal of nutrition, so it is better to blend your foods.

3. If you need the recipe to dehydrate faster, you can heat it up in a pot over medium heat for 10-15 minutes, stirring occasionally.

4. Spread your mixture out onto dehydrator trays that have been lined with tray liners or parchment paper. Thicken your puree around the edges of the tray, as they will dry faster. You should have about ¼" thickness at the edges of your trays and 1/8" thickness in the centers.

5. Dry your fruit roll-ups for 6 to 8 hours at 145F.

6. Roll your dried fruit roll-ups tightly and then cut them with a sharp knife. Wrap each piece with saran wrap and store it in an airtight container for up to 6 months.

Chapter 8

Freezing

Freezing is exactly what it sounds like: the type of food preservation that you engage in when you freeze foods. Virtually every household with a standard fridge has a freezer, which means you likely already engage in freezing foods at this time. Despite how common freezers and frozen food are, many people do not know how to use their freezers for the average ingredient. Instead, your freezer may typically be full of ingredients that you already bought frozen from the store, and you simply transferred them into your freezer.

Despite what you may think, not all things can just be tossed into the freezer. While most things are safe to be frozen, preparation methods must be followed to prepare those items to be frozen properly. Not following the proper freezing methods can lead to freezer burn, which can, in turn, reduce the quality of the food and drastically deteriorate the taste of the food, too. Each food type will have its own means for properly freezing said food.

What Foods Does Freezing Work for?

Most foods can be frozen, so it is actually easier to discuss what *can't* be frozen. Fresh tomatoes should never be frozen, though they can be cooked into sauces or pastes and then frozen that way. Whole eggs, rice, pasta, fried foods, salad greens, herbs, most sauces, cucumbers, and potatoes should never be frozen. While dairy products such as sour cream, yogurt, and milk may be frozen, know that the taste and texture will drastically change, so you will have to use them inside of a recipe if you are going to freeze them. Previously frozen meat should also not be refrozen. Although it will be safe to eat, it will seriously destroy the taste and texture of the meat.

What Materials Do You Need for Freezing?

Which materials you need for freezing depends on what it is that you are freezing. The following list of materials will provide you with everything you need to freeze all different types of foods, no matter what they are.

They include:

- Heavy-duty freezer bags
- Freezer-safe containers
- A large pot for boiling water
- A large bowl of ice water
- Tea towels for drying
- A pizza tray with holes in it
- A cutting board
- A sharp knife

How Does the Freezing Process Work?

The freezing process changes depending on what it is that you are freezing. When you are freezing things like meat or dairy products, you can freeze them by simply placing them in heavy-duty freezer bags or freezer-safe containers and freezing them that way. The same goes for most soups and sauces.

If you are going to be freezing fruits or vegetables with high water content, such as peppers, you are going to want to flash freeze them. You can flash freeze things by cutting them into the size you will want them to be in when you are consuming or using them later. Then, you will lay them on a pizza tray in a single layer, taking care not to let them touch each other too much.

Pop the pizza tray into the freezer and let the ingredients freeze for 1-2 hours, or until they are completely frozen through. Then, remove them from the tray and place them in a heavy-duty freezer bag or a freezer-safe container. This way, they do not freeze together in one big lump.

Other types of vegetables, such as carrots, broccoli, or cauliflower, should be blanched before freezing. To blanch vegetables, you will boil a large pot of water, while also preparing an ice bath for the vegetables you are blanching. You will then cut the vegetables up into the sizes you want to freeze them in and place them in the boiling water for up to 1 minute. Choose a few as 15-20 seconds for vegetables that are already softer, such as broccoli or cauliflower. Immediately transfer the vegetables to the ice bath. Once they are completely cool, transfer them to a tea towel to dry. Then, follow the flash freezing method to ensure that they freeze separately and not in one big lump. When they are done, transfer them into a heavy-duty freezer bag or a freezer-safe container.

Frozen Green Beans, Summer Squash, and Tomatoes

Makes as much as you would like.

What You Need:

- Green beans, ends removed

- Summer squash, peeled and chopped

- Tomatoes

How to Make It:

1. First, prepare your green beans. Chop off the stems on the ends, then cut your beans into whatever lengths you want. Lay a piece of wax paper over a cookie sheet and spread the beans out in a single layer. Place the entire cookie sheet into your freezer

and freeze the beans for 30 to 45 minutes, or until they feel frozen to the touch. Add them to your frozen mix bag.

2. Next, prepare your summer squash. To do this, you want to clean your squash first. Then, you want to slice it into rounds. If you want, you can also slice the rounds into quarters to create smaller pieces. Boil a pot of water and prepare an ice bath in a separate container. Once the water is boiling, add the squash to the boiling water and leave them there for three minutes. Begin counting as soon as you have placed your squash in the pot. When the three minutes are up, transfer them to the ice bath for five minutes. Then, drain them, dry them, and lay them on your flash-freezing tray and freeze them for about 30-45 minutes. When they are done, add them to your frozen mix bag.

3. Finally, prepare your tomatoes. To do this, wash and dry your tomatoes. Remove the stem and the core, then cut them into whatever size you desire. Place them on the flash freezing tray, skin side down. Freeze them for 30-45 minutes, or until they are frozen to the touch. Add your tomatoes to your frozen mix bag.

4. As long as you flash freeze all of your ingredients before adding them to your storage bag, they will not stick together too badly when you are ready to remove them from the bag to cook them. Be sure to keep them in a heavy-duty freezer-rated bag or container to avoid having your vegetables sustain freezer burn, as this will deteriorate the quality of your vegetables by damaging both their taste and their texture.

5. Eat your vegetables within 6-12 months for the best quality.

Frozen Meat

Makes as much as you would like.

What You Need:

- Meat

- Plastic wrap

- Aluminum foil

- Freezer-friendly zip-top bag

How to Make It:

1. Keep your meat in the package in which you purchased it. Tightly wrap the entire package in plastic wrap, then in aluminum foil. Place the entire thing into a freezer-friendly zip-top bag and store it in your freezer for up to three months. After 3-6 months, the meat will still be edible. However, the taste will begin to deteriorate rapidly.

Frozen Bread

1 loaf, or as many as you would like.

What You Need:

- A loaf of bread

- Plastic wrap

- Aluminum foil

- Freezer-friendly zip-top bag

How to Make It:

1. Remove your bread from its original bag. Or, if you made your own bread, keep it unpackaged. Cover the loaf in plastic wrap, ensuring that it hugs the edges of the bread tightly so that no air can get to it. Wrap it in one more layer of plastic wrap. Wrap the

entire thing in aluminum foil, then place it into a large freezer-friendly zip-top bag. If you do not have a large enough bag, wrap it in a second layer of aluminum foil. Freeze the bread for up to two months.

Chapter 9

Brining and Salting

Brining and salting are actually two different methods of the same preservation style. When brining, you use a wet brine to cure your meats, while salting means you use a dry brine to cure your meats. Wet brines are made of salt, sugar, and water and are accomplished by submerging the food item of choice in the brine with a heavy crock over the food item to keep it submerged in the brine. Dry brines are made of salt alone. They work by completely dusting your food in salt, which draws out the moisture and cures the meat. The trouble with dry brining, or salting, is that you might not be able to fully cover the meat in salt, which results in some of it spoiling. Further, if you add any additional salt, your food quality will be drastically reduced. For this reason, brining is the preferred method over salting these days, although both are still important to know about.

What Foods Does Brining and Salting Work for?

Brining and salting are mostly used with meats and seafood. Dry brining has also been used for vegetables like cabbage and runner beans.

What Materials Do You Need for Brining and Salting?

The tools you need to brine or salt food depend on if you want to brine or salt your food. Obviously, right? While not many tools are needed, it is important that you have the right ones to keep your food properly protected.

To brine your food with a wet brine, you will need:

- A brining bucket
- Brining salt

- A meat needle/pump

- A heavy crock

To salt your food with a dry brine, you will need:

- ½ TSP Kosher salt per pound of meat

- Fridge

How Does the Brining and Salting Process Work?

The brining process, or wet brining, works by combining brining sodium with water and pouring it in a brining bucket. Some brines will also call for a small amount of sugar to be added to the brine before adding it into the bucket. If you are in the wilderness or near the ocean, ocean water can be used for brining. However, you will need to bring it to a full rolling boil for about 5 minutes and then let it cool completely to kill off any unwanted bacteria and parasites.

Once the brine is prepared, you will pour it into the brining bucket. Then, you will take your meat needle or pump and draw brine up into the meat needle or pump, then inject it into the meat itself. Next, you will place the meat into the brining bucket and use a heavy crock to weight it down. You need to make sure the meat is completely submerged into the brine so that it is properly preserved. If any of the meat is above the surface of the brine, it will spoil.

For salting, or dry brining, you will rub the salt into the surface of the meat. Massage it until the meat is completely covered, and the salt sticks easily and evenly across the surface of the meat. You will need to leave it for 2-4 hours for a thinner cut or 2-4 hours for a larger one. Particularly thick cuts, such as poultry breasts, should cure for 4-6 hours. Roasts should cure for 12-48 hours. They can be kept in the fridge until you are ready to use them.

Brined Pork

4 Pork Chops

What You Need:

- 3 cups water

- ¼ cup firmly packed brown sugar (golden)

- ¼ cup pickling salt

- 2 cups ice cubes

- 4 pork chops

How to Make It:

1. Dissolve water, salt, and sugar. If you need to, you can do this over low heat on the stove, so everything dissolves faster. Add ice to your brine. Let the brine cool to lower than 45F.

2. In a zip-top bag, place 4 pork chops and cover them with brine. Place in the fridge for 2-6 hours, then use them in your desired recipe. If you are not ready to use them, remove them from the brine and store them for up to 2 days in plastic wrap in your fridge.

Brined Cheese

1 block of cheese.

What You Need:

- 2.25lb salt

- 1-gallon water

- 1 TBSP calcium chloride (30% solution)

- 1 TSP white vinegar

How to Make It:

1. Combine brine ingredients.

2. Use the freshest cheese you have, or make your own. Ensure it is cooled before placing it into your brine. Salt the top surface of the cheese, as the cheese will float, and the top surface will not be properly coated. Halfway through the brining process, flip the cheese and salt the top once again to protect it. For hard cheeses, you will need to brine them for 24 hours. For softer cheeses, they will require 12 hours of brine.

3. After the required amount of time, remove your cheese from the brine and allow it to air dry. It should take 1-3 days for the air-drying process. You will know that it is done when you can see a firm, dry surface on your cheese. At this point, you can wax your cheese or allow it to develop a natural rind. If you allow it to develop a natural rind, be sure to find instructions on how to do so with the specific cheese you are using, to avoid accidentally causing spoilage in your cheese.

Chapter 10

Sugaring

Sugaring is similar to pickling; however, it works a little differently. To sugar something, you will first dehydrate it; then, you will pack it with sugar. The sugar creates a hostile environment for bacteria, which prevents spoilage. Sugaring can be used in a pinch, and for some things, it is great. However, you must be cautious with this method as sugar itself can attract moisture, which, as you know, can spoil foods. If moisture rises in the sugar itself, native yeast in the sugar and the environment will come out of dormancy and begin to ferment the sugars. While fermentation is a form of food preservation, when it is accomplished by accident, it can lead to unpleasant situations where the food itself would taste gross. It could lead to you falling ill if you consume it. Following exact methods when sugaring food and storing sugared food is the best way to make sure you are not putting yourself at risk of getting sick from sugared foods.

What Foods Does Sugaring Work for?

Sugaring is usually used with fruits; however, certain vegetables such as ginger can be stored using the sugaring process.

What Materials Do You Need for Sugaring?

Sugaring food is done by completely dehydrating foods, than by cooking them in liquid sugar. You then remove the item from the sugar syrup and allow them to dry completely before storing them in airtight containers.

To complete the sugaring process properly, you will need:

- Fully dehydrated food
- A large pot

- Sugar

- Water

- Airtight containers or a vacuum sealer

How Does the Sugaring Process Work?

The sugaring process works using any raw sugar or table sugar that is in crystallized form. However, syrup, honey, and molasses can all be used to store foods this way, too. Before you begin the preservation process of a fruit or vegetable, you will thoroughly wash it and then dehydrate it completely. Then, you will cook the dehydrated food in liquid sugar until the food itself becomes crystallized. Once it has, you can remove the food from the sugar, dry it, and preserve it in airtight containers or bags until you are ready to use them. Because of how finicky sugared food can be, a vacuum sealer is generally the best way to preserve your sugared foods as it will ensure that no moisture can get into your final product.

Sugar Preserved Citrus Fruits

1 Pint Jar.

What You Need:

- Citrus (make a mix of your favorite citrus fruits such as oranges, limes, lemons, etc.)
- Granulated sugar

How to Make It:

1. Quarter your fruit and place a layer into a sterilized canning jar. Sprinkle with one tablespoon of sugar. Repeat this until the entire jar is full. Then press down on the fruits, which will release some juices into the bottom of the jar.

2. Cover the entire contents of the jar in 2-3 more tablespoons of sugar, as much as you can fit.

3. Place the jar in the fridge for three weeks so it can "rest."

4. Eat your sugared fruits anywhere from 3 weeks to up to 6 months.

Chapter 11

Smoking

Smoking is a type of food preservation method that results in foods tasting similar to barbecued dishes. The difference with smoking is that it has a delicious, naturally smoky flavor based on the type of preservation method used and that it can be preserved for much longer, and much easier than a barbecued dish can be. While a barbecued dish must be stored in the fridge and eaten within a few days, smoked foods can be vacuum sealed and eaten within a few months or up to a year.

What Foods Does Smoking Work for?

Meats and fish are the most common foods to be smoked, though you can also smoke cheese and certain vegetables. Some ingredients that are used in beverage making are smoked, as well, to give beverages a nice smoky flavor. Since you are preserving food for your family, it is most likely that you will be smoking different meats and fish.

What Materials Do You Need for Smoking?

In order to start smoking your foods, there are a few different tools you are going to need. A smoker, fuel, and wood chips are the first tools you will need. You will also need utensils for placing and removing your smoked preserves, and meat or fish for you to smoke.

Types of Grills

There are two types of grills you can use when it comes to smoking meats. Your classic outdoor barbecue grill is one you can use, or you can purchase a proper smoker. If you are going to be smoking a lot of meats, it is better to buy a smoker as it will do a much better job of smoking the foods you will be preparing.

If you are using a grill, you are going to need to be able to create enough smoke to cook your meats effectively.

On a charcoal grill, you can do this by placing coals on one side of the barbecue and placing a drip pan on the other side. Once your coals are hot enough to cook over, place a layer of wood pellets over them. If you place a layer of liquid in your drip pan, you will get much better smoking results this way. Apple juice is a great liquid to use here as it will add flavor to your meat during the smoking process. When you are ready to start cooking, place your meat over the drip pan. Be sure to keep a small opening in your lid so that the smoke can vent out properly. This keeps smoke moving around the meat so that it can properly penetrate the surface and cook it through.

If you are using a gas grill, there are no charcoals for you to lay the wood over. In this case, you need to place your wood chips in a metal pan, which will be inserted directly over the gas flames. Set this pan to one side. Let the entire setup preheat for about twenty minutes before placing your meat on the side opposite where the wood chips were wet up. Then, close the lid leaving space for ventilation to occur during the smoking process.

Types of Wood

There are many different types of wood that can be used with smoking recipes. The type of wood you choose will affect the flavor of your meat, so you want to pick one that is going to give you the best flavor. Avoid using softwoods like pine or cedar, as these will produce too much smoke and will release a resin that will penetrate your meat and create a resinous flavor.

The types of wood that are best to use with smoking, and what types of meat they can be used for, can be found here:

- Alder: Chicken, seafood, pork.

- Apple: Chicken, seafood.

- Cherry: Chicken, seafood, pork, beef.

- Hickory: Pork, beef.

- Maple: Chicken.

- Mesquite: Pork, beef.

- Mulberry: Chicken, seafood, pork.

- Oak: Chicken, seafood, pork, beef.

- Peach: Chicken, pork.

- Pear: Chicken, pork.

- Pecan: Chicken, pork, beef.

- Walnut: Pork, beef.

While you can certainly use any of these woods with any meat you like, they are unlikely to taste good with meats that are not listed here. Chicken and seafood tend to pair well with sweeter, lighter woods since they absorb more flavor, which means heavier woods can become too strong in these meats. Alternatively, pork and beef do not absorb quite as much of the flavor, which means that you can go ahead with stronger and heavier woods because they will penetrate the meat and leave a better flavor in the end.

When you purchase your meats, you will notice that there are many options. Wood chunks, chips, and pellets are all offered as wood sources for smokers. For longevity and sustainability, chunks and chips are the only types of wood you should really consider. While pellets will work, they tend to burn off quickly, which will leave you having to add more far more frequently than you would with chunks or chips.

Wood chunks are excellent if you are going to be smoking food all day long. If you are going to be preserving large amounts of food with your smoker, these chunks will save you from having to add more wood to the smoker as often. This prevents you from opening the smoker too frequently, meaning your food cooks better. To get even more smoke out of your wood chunks, you can soak them in water for up to one hour before you begin smoking your food so that they are nice and moist, as this will cause them to smoke more.

Wood chips are great if you are only going to be smoking food for a couple of hours at a time. These are smaller than the chunks, and so they will burn faster, though that is not a problem if you will not be using your smoker all day. The best way to get the most out of your chips is to soak them in water for up to 30 minutes before you start so that they produce more smoke.

How Does the Smoking Process Work?

Smoking your food requires you to provide your food with indirect heat over lower cooking temperatures and with an abundance of smoke present. The smoke itself creates an acidic coating on the surface of the food, which prevents bacteria from growing on or in the meat. It also dehydrates the food, which creates an environment that is less hospitable for bacteria to grow. Smoking also enhances the flavor of the food you are preparing, making it an excellent option for preserving foods. Smoked foods can be served as is, or they can be placed in vacuum-sealed bags and stored for longer periods of time. The exact amount of time a food can be stored depends on what that food is and how well it was smoked.

There are two types of smoking: hot smoking and cold smoking. Hot smoking happens at temperatures over 150F, while cold smoking happens at temperatures less than 100F. Hot smoking is used for cooking the food while also providing a rich smoky flavor, while cold smoking will only flavor the food, it will not cook it. Certain already-cured meats such as salami and cheeses such as cheddar are cold smoked as a way to enhance their flavor.

The first step of any smoking recipe will always be to prepare your food first. For meats, this requires you to trim the meat into the desired shape and then marinade it in a dry brine, which will help cure the meat. You will store the meat with the dry brine in the fridge for up to five days, flipping it each day to ensure that the cure evenly distributes across the meat. Once five days have passed, you will rinse the meat under cold water and allow it to dry out in the fridge. The next day you will prepare your smoker by adding the necessary wood chips or chunks and allowing it to preheat for about 20 minutes, as this will ensure that it is nice and hot and the smoke is abundant for the meat. Then, you will place the meat in your smoker. Each type of meat will require different smoke times, so you will need to follow a specific recipe for your desired ingredient.

By following this entire process, you effectively prepare the meat for smoking by helping to first dry it out using the dry brine. Then, come the smoking day, you allow a combination of heat and smoke to properly cook the meat while also creating an acidic environment on the meat which prevents it from going bad. After the entire smoking process, the meat will be dehydrated enough that it will be able to be preserved for long periods of time. The key to success here, though, is to realize that the meat will not be completely free of moisture, which means bacteria *can* still grow on it. A proper vacuum-sealed bag will prevent the moisture content from increasing or spreading and will ensure that you can enjoy your meat for months to come.

Smoked Pastrami

Five to six pounds of pastrami, depending on the size of the meat you cook.

What You Need:

- 1 brisket (5-6 pounds)

- 1 cup + ½ cup brown sugar

- ¾ cup kosher salt

- 2 TBSP mustard seed, whole

- 2 TBSP paprika

- 1/3 cup + 2 TBSP coriander seed, whole

- 1/3 cup + 2 TBSP black peppercorns, whole

- 1 TBSP ground ginger

- 1 TBSP red pepper flakes

- 1 TBSP allspice berries, whole

- 1 TBSP granulated garlic (not garlic salt)

- 1 TBSP granulated onion (not onion salt)

- 2 TSP pink curing salt

- 6 cloves, whole

- 4 garlic cloves, whole

- 6 bay leaves

- 1 gallon filtered water

How to Make It:

1. Trim the fat off the top of the brisket.

2. Pour 1-gallon water into a large, non-reactive container. Add 1 cup brown sugar, ¾ cup kosher salt, 2 tbsp whole mustard seed, 2 tbsp whole coriander seed, 2 tbsp whole black peppercorns, 1 tbsp ground ginger, 1 tbsp allspice berries, 2 tsp pink curing salt, 6 whole cloves, 4 whole garlic cloves (peeled,) and 6 bay leaves.

3. Wrap the brine container and cure the brisket for 7-10 days, flipping it over daily to ensure that it cures evenly.

4. On cook day, preheat your smoker to 225-250F. Add your desired wood chips or chunks to the coals. Close your smoker and let the smoke begin to develop inside.

5. Prepare your dry rub by combining ½ cup brown sugar, 2 tbsp paprika, 1/3 cup whole coriander, 1/3 cup whole black peppercorns, 1 tbsp red pepper flakes, 1 tbsp granulated garlic, and 1 tbsp granulated onion.

6. Thoroughly rinse the brisket under cool water, then pat it so that it is completely dried. Add the dry rub and massage it into all sides of the brisket, applying liberally as you go.

7. Place your brisket in the smoker and let it smoke until it reaches 150F inside. This should take around 3-5 hours.

8. Preheat your oven to 250F. Pour water into a roasting pan so that it is 1" deep, and place a rack over it so that the pastrami is not sitting in water. Add an oven-safe probe into the pastrami, place it on the rack, and cover it with tinfoil. Cook until your pastrami reaches 230F inside. This should take around 4-5 hours.

9. Slice your pastrami and serve it, or store it in vacuum-sealed freezer bags for up to one year.

Maple Smoked Bacon

3 pounds of bacon.

What You Need:

- 3lb unsliced pork belly, about 1" thick and 6-8" long.

- 3 TBSP dark brown sugar

- 3 TSP black pepper, ground

- 3 TBSP pink curing salt

- 3 TBSP maple syrup

How to Make It:

1. Combine dark brown sugar, black pepper, pink curing salt, and maple syrup in a non-reactive container filled with about ½ gallon of filtered water. Add the pork belly to the container and let it sit for seven days, turning it daily for even curing.

2. After the 7th day, completely rinse the pork belly and let it sit in the fridge on a wire cooling rack for 12-24 hours, until it develops a sticky skin. This sticky skin is called "pellicle."

3. Prepare your wood pellets in your smoking machine and set the temperature to 165F. Let it preheat for around 20 minutes. Add your pork belly and smoke it for about 6 hours, or until it is 155F inside.

4. Slice the bacon. Store it in vacuum-sealed bags in the fridge for up to 2 weeks or in the freezer for up to 1 year. When you want to cook it, defrost it and fry it in a pan as usual.

Smoked Jerky, Black Pepper Flavor

4-6 Servings of smoked jerky.

What You Need:

- 1 bottle dark beer

- 1 cup soy sauce

- ¼ cup Worcestershire sauce

- 4 tbsp black pepper, ground and divided

- 3 tbsp brown sugar

- 1 tbsp pink curing salt

- ½ tsp garlic salt

- 2 pound trimmed beef top, sirloin tip, bottom round, flank steak, or wild game

How to Make It:

1. Make a marinade from beer, soy sauce, Worcestershire sauce, 2 tbsp black pepper, brown sugar, curing salt, and garlic salt.

2. Trim fat away from your meat and slice it into ¼" thick strips, against the grain. If it is too hard to slice your meat, place it in the freezer for about 30 minutes, and try again. This should firm up the meat, so it is easier for you to cut.

3. Place beef slices in a large zip-top bag and fill it with the marinade. Massage the bag to penetrate the meat with the marinade and to ensure that everything is evenly covered. Seal the bag and place it in the fridge overnight, or for up to 24 hours.

4. Prepare your smoker with a temperature of 180F. Remove beef slices from marinade and dry them between paper towels. Sprinkle the dried slices with the remaining 2 tbsp black pepper. Discard the remaining marinade.

5. Lay beef jerky strips in a single layer on the grill grate. Smoke for 4-5 hours, or until the jerky is chewy. It should be somewhat pliant when you attempt to bend a piece of your jerky.

6. When the jerky is done, transfer it to a zip-top bag while it is still warm, and let it rest on the counter for one hour at room temperature. Seal the bag, then place it in the fridge for up to 2 months. Or, vacuum seal the bags and place them in the freezer for up to 7-12 months.

Smoked Fish

One full fish, smoked.

What You Need:

- 5 pounds salmon, trout, or char

- Maple syrup

- 1 quart filtered water, cool

- 1 cup brown sugar

- 1/3 cup kosher salt

How to Make It:

1. Begin by curing your fish. Combine water, brown sugar, and salt in a non-reactive container and place the fish in it. Put it in the fridge for at least 4 hours, but up to 48 hours. Never go over 48 hours, or your fish will be too salty. Fish is softer than other meats, so it absorbs the liquid much faster. Flip the fish halfway through your curing time for even curing.

2. Remove the fish from the brine and pat it dry without rinsing the fish. Place the fish on a cooling rack, skin side down, and let them cool at 60F or cooler for 2-4 hours. Alternatively, you can cool it overnight in your fridge.

3. Use soil to lightly coat the skin of your fish, as this prevents it from sticking to the rack in your smoker. Then, place the fish on your smoker's grill, skin side down. Smoke your fish between 140F and 150F for an hour, and then finish it at 175F for 1-2 hours, or until the fish is cooked through.

4. Every hour that your fish is in the smoker baste it with maple syrup. This will remove any albumin that may form so that your fish stays tasty and fresh. Avoid letting the heat get too high, as this will result in more albumin and can overcook your fish.

5. Once the fish is done, cook it to room temperature for one hour, then wrap it in an airtight container and place it in the fridge for up to 10 days. Or, you can vacuum seal your fish and store it in the freezer for up to 1 year.

Chapter 12

Pickling and Fermenting

Pickling and fermenting are two types of preservation methods that use salt brine to create a delicious, sour outcome. Pickles themselves are something we are likely all familiar with; however, you may not be familiar with fermented foods and the taste of fermented ingredients. A great example of fermented foods that you can buy at the grocery store includes sauerkraut, horseradish, and kimchi. When you ferment food, you create an environment for that food through the help of salt brine and temperature that allows it to begin the fermentation process in a controlled manner. The result of properly fermented foods is the creation of food that is rich in probiotics, and that promotes healthy gut flora.

Fermentation and pickling may sound similar, but there are differences. With pickling, you are soaking food in acidic liquids to achieve a sour flavor, and preserving that food happens by proxy. Most pickled foods are pickled using a vinegar brine. With fermented foods, the sour flavor you taste is the result of a chemical reaction that occurs within the food between the natural sugars in the food and naturally present bacteria. You do not need acidic liquids to create a fermented food. Most fermented foods are fermented using a salt brine. You can make fermented pickles, as well, which is a unique recipe that combines the properties of both of these preservation methods to create a delicious result.

What Foods Does Pickling Work for?

There are many different types of food that you can pickle. The most obvious is cucumber, though you can also pickle asparagus, beets, bell peppers, cauliflower, carrots, green beans, onions, peaches, radishes, turnips, and even certain fish such as herring.

What Materials Do You Need for Pickling?

To properly pickle any food, all you need is pickling salt, vinegar, glass jars with proper sealing lids, vegetables or fruits, and herbs. You will also need a nice dark place to store your pickles as they absorb the liquids, *or* you can store them in your fridge if you are making fridge pickles. Fridge pickles will take up more space in your fridge. However, they will last much longer than fresh vegetables being stored in your fridge, so this does technically preserve them for a more significant period of time.

How Does the Pickling Process Work?

Pickling with vinegar brine works by using a simple mixture of water, salt, and vinegar to create a brine that effectively pickles your foods. There are many sub-methods of pickling, including pickling methods that allow for different flavors and that work for different types of foods. The distinction between these methods generally includes using different herbs or different ratios in the vinegar brine to create your desired flavor. Pickles can either be water bath canned so that they can be preserved at room temperature, or they can be made in the fridge, which means they are not sealed or shelf-stable, but they do last much longer than fresh vegetables would. Fridge pickles tend to mature faster than shelf-stable pickles, so they both have their own set of benefits.

As a food item sits in the vinegar brine, it absorbs the brine, which gives it a nice, sour flavor. This flavor will directly reflect the brine you used and anything that was placed in the brine with your pickles. This is how you make the flavors present in garlic pickles, dill pickles, spicy pickles, and other flavors that you might come across.

Cucumber Carrot Fridge Pickles Recipe

2 x 250mL Jars of Pickles

What You Need:

- 3 cups water

- ¼ cup rice wine vinegar

- 2 TBSP sugar

- 1 tbsp + 1 tsp kosher salt

- 1 cucumber, thinly sliced

- 2 shallots, thinly sliced and separated into rings

- 1 carrot, thinly sliced

- 1 red Thai chile, stem removed and thinly sliced

How to Make It:

1. Boil 3 cups of water. In a bowl, combine cucumber, shallots, and carrot. Cover the produce in 1 tbsp salt. Ensure the produce is in a heat-proof bowl. Pour boiling water over the picture. Stir, and let stand for 20 minutes.

2. Drain vegetables in a colander until they are completely drained. Combine vinegar, sugar, chile, and 1 tsp salt. Dissolve the sugar and salt completely. Divide between 2 x 250mL jars, and then divide the produce between the jars, too. Place in the fridge for at least 1 hour. The longer they sit, the better they will taste. They can be kept for up to 1 year.

Pickled Egg Recipe

12 eggs.

What You Need:

- 12 eggs

- 1 cup white vinegar

- ½ cup filtered water

- 2 TBSP coarse pickling salt

- 2 TBSP pickling spice

- 1 onion, thinly sliced

- 5 black peppercorns

How to Make It:

1. First, hard boil your eggs by placing them in a pot and covering them in cold water. Bring the pot to a boil and immediately remove the pot from the burner, letting it stand for 10-12 minutes. Cool the eggs in a cool bowl of water, then peel the shells off. Place them in a 1-quart wide-mouthed jar, or divide them between two half quart wide-mouthed jars.

2. Combine vinegar, water, salt, and pickling spice in a saucepan. Add most of the onion, except for a few slices, and the five black peppercorns. Boil the mixture until it reaches a rolling boil, then pour the mixture over the eggs. Add the reserved onion slices on top and cool to room temperature. Place in the fridge for three days before serving and eat within a year. Or, use a water bath canner to preserve the eggs for a longer period of time.

What Foods Does Fermenting Work for?

Fermenting foods is most common for vegetables and fruits. The fermentation process can also be used to make wine, as well as wild-caught yeast in the form of a sourdough starter. Fermentation happens when you allow the natural sugars present in an ingredient to engage in the fermentation process in an intentional, controlled environment. The best foods to ferment include, cabbage, carrots, cauliflower, cucumber, garlic, kohlrabi, pepper, radish, snap beans, and turnips. Foods with high water content work best for fermentation.

What Materials Do You Need for Fermenting?

Fermenting requires you to have a fermenting crock or jar, vegetables, herbs, pickling salt, plastic bags, and water. You will also need a sharp knife to fine-cut your fermented foods, as thicker foods do not ferment as well as thin-sliced foods.

How Does the Fermenting Process Work?

The fermenting process works by taking salt and massaging it into the vegetable of choice until it begins to wilt, and water begins to be extracted from the vegetable. It should take around three to five minutes to massage your vegetables to the point where they are ready to ferment. Once you have effectively massaged them, you will stuff them firmly, but not too tightly, into mason jars. If you have proper food-grade pickling crock, you can do this directly in the crock and then simply leave the ferment in the crock. If you do not have enough liquid to completely cover the food that you are fermenting, you will need to add a saltwater brine to it. Then, you will place a weight over the food to ensure that it stays under the brine. Unless you have a proper crock set up, the easiest way to do this is to insert a (clean) plastic bag into your mason jar. Then, fill it with the remaining brine, or with fresh filtered water, and seal it. This should create a closed seal over the top of the food, keeping it submerged in water and protected from air. It is important that fermented foods are kept in air-tight containers to avoid having air access the fermenting food. Air can increase the likelihood of mold forming or spoilage occurring before the fermentation process has a chance to really get started. Each day you will stir the fermented food and then replace the water bag into the jar to create the airtight seal again. You will want to keep the jar on a plate or in a small bowl, as it will "burp" air through the sides of the bag, which will bring some liquid with it, too. Most fermented foods are finished anywhere from a week to four weeks, depending on what flavor you are looking for. Once you reach your desired flavor, you will place your ferment into sterilized mason jars, leaving ¼" headspace. You can then put it in the fridge or water bath can it, depending on what your unique recipe calls for. If you place your ferment into the fridge, the fermentation process will be extremely slowed down to the point where you will not likely notice it.

The only type of fermentation that does not include salt is a sourdough starter. Sourdough starter is technically a fermented food, though it does not require salt, and it cannot be eaten as is. With sourdough, you do a mixture of flour and water on your counter and continue to "feed" it with more flour and water every single day. This results in the starter capturing wild yeast, meaning you do not have to use active yeast to create your sourdough bread. This starter can then be incorporated into sourdough recipes such as bread, pancakes, pizza crust, pretzels, cinnamon buns, and more. It is said that sourdough is easier to digest because of the fermentation factor, which makes this a great healthy bread option and also makes for an excellent alternative to store-bought yeast. During the 2020 pandemic, this is a wonderful alternative if you find that your local store is sold out of conventional yeast.

Traditional Fermented Sauerkraut

Makes about 1 to 1.5 quarts of sauerkraut.

What You Need:

- 1 medium head green cabbage
- 1.5 tbsp kosher salt

How to Make It:

1. Completely clean all of your preparation tools, first. This way, your ferment has the best chance at developing healthy, clean bacteria, rather than the bad kind that will cause spoilage.

2. Remove the outer leaves of cabbage. Quarter it and cut out the core. Slice wedges into thin ribbons of cabbage and place them in a large, non-reactive bowl. Sprinkle 1.5 tablespoons of salt over your cabbage and start massaging it with your clean hands. As you massage it, the cabbage will begin to wilt, and water will begin to pour out of the leaves. It will start to look like coleslaw, rather than raw cabbage.

3. Pack sauerkraut into clean 500mL mason jars. It should take about three mason jars, as you want to leave at least 1.5-2 inches of headspace in each jar. Pour the liquid released by the cabbage out of the bowl and into the jars. If the liquid does not cover the cabbage, combine 1 tsp of salt with 1 cup of water until the salt dissolves, then pour that into your jars to cover the cabbage.

4. Weigh the cabbage down either using a smaller jelly jar full of clean stones or marbles or using a bag full of excess brine or filtered water. If using a jar, cover the entire thing with a cloth and use a rubber band to secure it in place. If using a bag, place a clean bag inside of the jar over top of your sauerkraut and use your hand to lightly press the bag out from the inside of the bag itself. Then, fill it with brine until the brine reaches the mouth of the jar. Remove all air from the bag and seal it, then move it around a bit to release any air bubbles that may have become trapped under the bag.

5. Over the next 24 hours, press the cabbage down about 3-4 times so that it becomes even more wilted.

6. Let the sauerkraut ferment between about 65F and 75F for 3-10 days. At temperatures over 75F, your sauerkraut will ferment rapidly and may become fairly soft in the

process. Below 65F, your sauerkraut will not have enough warmth to encourage the growth of the good bacteria in the ferment.

7. As your sauerkraut begins to come "mature," you can start tasting it. Early on, it will have a salty taste. Over time, it should develop a sour, tangy scent, and taste, which indicates that it is done. If you see scum starting to develop on the top of your sauerkraut, that is normal. Skim it off and discard it. If you see any mold, discard your entire jar and start over, as mold can make any cabbage that touched the mold dangerous to eat, and it can damage the taste of the rest of your sauerkraut.

8. Your complete sauerkraut can be transferred into either 4 x 250mL jars or 2 x 500mL jars once it is complete. Then, store it in the fridge. It can be eaten for up to 2 months, though it will typically last much longer. So long as it smells good to eat, and tastes good to you, the sauerkraut should still be fine to eat. Ferments like sauerkraut will rarely spoil, especially when kept in the fridge; however, they can begin to taste off over time.

Fermented Beet Slaw Recipe

Makes about 1 quart.

What You Need:

- 6 cups raw beets, peeled and shredded
- 1 tbsp kosher salt

How to Make It:

1. Lay shredded beets in a mixing bowl and sprinkle the salt over them. Let them sit for about 5 minutes, then use your clean hands to massage the beets to ensure that they all

absorb the salt. As you do this, the shredded beat pieces will become wilted and will expel even more water.

2. Pack the beets into a clean, quart-sized jar and leave at least 1 inch of headspace. This way, the beets can safely release juices and gasses during the fermentation process. You will need to "burp" the jar every so often to release the gasses so that the jar does not explode. You should burp your jar 2-3 times a day, or more if you notice that it is becoming particularly concentrated.

3. Ensure that the lid of your slaw is airtight to avoid any air getting into your ferment.

4. After 2-3 days, the beet slaw will be done. Refrigerate it for up to 2 months.

Chapter 13

Ash, Oil, Honey

Ash, oil, and honey are all used as a means for preserving foods. Each one of them has its own unique benefits and capabilities. They also share the common qualities of being lesser-known preservation methods that have stood the test of time and offered great preservation solutions to people for many generations. With all three of these preservation methods, the main goal is to coat the food you are preserving so that no air can access it. They also have their own antibacterial properties, which enable the foods within them to remain safely stored and able to be consumed for long periods.

What Foods Does Ash Work for?

Ash is an incredibly traditional food preservation method that can be used to store things like eggs, cheese, and tomatoes.

What Materials Do You Need for Ash?

In order to effectively preserve your food in ash, you will need fresh, sifted ash, an airtight container, and the food you intend on preserving.

How Does the Ash Process Work?

The process of storing your food in ashes is incredibly simple. You will take sifted ash and place a layer on the bottom of an airtight container. Then, you will layer your food into the ash, and then you will cover it with ash again. You will continue forming these layers until the box is full. Then, you will replace the lid on the box and store it this way for up to three months.

Ash Preserved Gruyere Cheese

Makes 1 block of ash-preserved cheese.

What You Need:

- 1 block of Gruyere, not a thin piece
- 3" of sifted wood ash

How to Make It:

1. Remove Gruyere from the fridge and let it dry off at room temperature for an hour or two so that it is not retaining too much moisture.

2. Cover the bottom of a stoneware pot in 1.5" of sifted wood ash. Place Gruyere cheese over the ash and cover it in another 1.5" of sifted ash. Store the pot of cheese in a cool root cellar for up to 3 months.

Ash Preserved Garden Fresh Tomatoes

Makes 15 tomatoes.

What You Need:

- 15 tomatoes
- 3" or more of sifted wood ash

How to Make It:

1. Pick tomatoes fresh from your garden. Choose ones that are ripe but not soft or overripe. Do not store tomatoes with bruises or blemishes, as it will not work. Line a wooden or cardboard box with paper, then place 1.5" of sifted wood ash in the bottom

of the box. Layer your tomatoes onto the ash, taking care not to let them touch each other. Cover them with more sifted wood ash until there is 1.5" of ash over the tomatoes. Place it in a cool, dry place, such as a root cellar, for up to 3 months.

2. This process will result in the skin of the tomatoes wrinkling somewhat over time. However, the inside will remain juicy and fresh the entire time.

What Foods Does Oil Work for?

Oil is best for preserving foods like sun-dried tomatoes, baby artichokes, sweet peppers, eggplants, mushrooms, garlic, goat cheese, basil, lemons, sardines, and tuna fish.

What Materials Do You Need for Oil?

To properly store food in oil, you will need a clean, airtight container, olive oil, and the food that will be stored in the oil.

How Does the Oil Process Work?

The process of storing food in oil is simple. You will place your chosen food into an airtight container and then cover it with oil. Then, you will place the lid on your food and allow it to sit. The oil prevents any air from accessing your food, which means it lasts longer. For certain foods, such as seafood, you will need to cure them with salt first so that they are ready to be preserved in oil. Otherwise, they will not be safe to eat, and they can contaminate the oil with unsafe materials, too.

Garlic Preserved in Olive Oil

About 15 cloves of garlic, or more as needed.

What You Need:

- 15 cloves of garlic, peeled

- 250mL of olive oil

How to Make It:

1. Place garlic in a 250mL mason jar. Pour oil into the jar until it has completely covered all of the garlic. Close the lid on the jar until it is finger tight. Place it in the fridge for up to 3 months. Replenish the olive oil as needed.

Fresh Tuna Preserved in Olive Oil

1lb tuna in oil.

What You Need:

- 1lb tuna

- 1.5 cups olive oil

- 1 garlic clove, crushed

- 3 sprigs thyme

How to Make It:

1. Rinse tuna thoroughly, then use a paper towel to pat it dry. Cut the tuna down into smaller 1" portions. Place the tuna in a small saucepan with 1.5 cups of olive oil.

2. Heat the tuna over low heat until small bubbles start to come to the surface. Let it cook for 10 minutes, without letting the oil come to a boil. You want it to maintain a nice, low simmer that will gently heat the tuna without deep frying it.

3. When the tuna is cooked through, add the garlic and thyme, then pour the contents into a one-quart mason jar and let it cool to room temperature. Cover the jar, place it

in the fridge, and let it infuse overnight before serving. You can save tuna this way for up to 2 weeks in the fridge.

What Foods Does Honey Work for?

Honey works best in specific canning recipes that call for honey, or with dehydrated fruits or herbs.

What Materials Do You Need for Honey?

To store things with honey, you will need raw, unpasteurized honey, a container for storing everything in, and dehydrated fruits or herbs.

How Does the Honey Process Work?

Using honey to store your food is a great alternative to using sugar. Honey can be used to replace sugar in certain canning recipes. Though, food can also be stored directly in honey since honey itself is antibacterial, and it is so thick that it will prevent any oxygen from getting to the food that is being preserved. In fact, this is why honey often makes for a great dressing for fresh wounds.

Aside from using honey to substitute sugar in canning recipes, foods can be stored directly in honey as well. To do so, you will dehydrate, or at least partially dehydrate, fruits, and then place them in honey. The honey would then keep them from fermenting, rotting, or spoiling for several months.

Honey Berry Puree

Makes 1 pint of honey berry puree

What You Need:

- 500g raw honey

- 160g strawberries

How to Make It:

1. Clean strawberries and remove the stems. Puree them in a blender without adding any water.

2. Mix the puree and honey together in a bowl. Pour the mixture into a pint-sized mason jar and seal. The honey will be shelf-stable as is, without canning it or putting it in the fridge. You should eat it within three months for freshness.

Chapter 14

Food Preservation Safety FAQ and Tips

As you likely already know, you have to be exceptionally careful with how you handle your food. When it comes to food preservation, the methods for keeping your food safe are slightly altered. In typical cooking, you use fresh food and heat it to specific temperatures to ensure safety. You may be familiar with storing your food in the fridge and in the freezer. However, these are two preservation methods that are generally easy and commonly accepted in our modern society. That means you have a lot of experience with safely storing your food in the fridge and freezer and preparing meals from your refrigerator and freezer.

When it comes to other methods of preserving food, especially methods that will be storing food at room temperature for any period, there are many things that you need to consider. These things will allow you to ensure that your food is preserved in optimal conditions so that you are confident that it is safe for you to consume those foods.

Understand that of all of the food preservation methods I have provided you within this book; each will offer you different lengths of preservation. Some will only extend the life of your food by a few weeks, while others will extend it by months or even years. This is just one factor that you need to be aware of when it comes to properly preserving your food so that you can create a hearty stockpile for your family. It is crucial that you consider best by dates and food spoilage in your plans when it comes to preserving foods, as you want to build a preservation schedule that will provide you with plenty to eat. Some things you may have to preserve multiple times throughout the year, while others may be safe to preserve only once per year or even once every other year. Naturally, the food items that can be preserved for longer periods can be preserved in larger batches since you have longer to consume everything. Foods that can only be preserved for short periods should be preserved in small batches so you can reasonably consume everything before it spoils. Although this may not leave you with an indefinite bounty of food, it will certainly allow you to stock more food for longer periods so that you are never without an abundance of things to eat in your pantry.

Each preservation method will have its own unique set of safety measures that should be taken into consideration when it comes to preserving food. I have included information on each of the methods written about in this book so you can feel confident in safely preserving your food every single time.

Standard Food Preservation Safety Considerations

When it comes to preserving foods, there are a few things that apply no matter what type of food preservation you are doing. Following these general standards, every time is imperative to the safety of your food. In many cases, not following these proper standards will not only reduce the safeness of your food, but it will also reduce the quality of the flavor of your food. For both of these reasons, it is imperative that you follow all of these considerations.

Keep Your Space, and Tools, Sterile

Whenever you are working with food, you should always keep your workspace sterile. When it comes to preserving food, however, you need to be even more cautious as you do not want to transfer bad bacteria into your preserves. Even a small amount of bad bacteria can rapidly multiply in a preserve recipe, leading to massive and dangerous spoilage over time.

Immediately before you begin preserving your foods, ensure that you clean everything in the hottest water possible and with proper antibacterial soaps and detergents. Keep your hands clean, as well, and prevent food from cross-contaminating other food sources. It is best to educate yourself on proper safe kitchen etiquette and follow these standards, particularly when you are preparing foods for preservation recipes.

Only Use Tools That Are In Proper Working Order

Never use any tool that is not in proper, safe working order. Most food preservation tools can be rather dangerous if they are not in proper working order, which can lead to injury, illness, or serious damage. For example, pressure canners that are not kept in proper working order can explode and damage the top of your oven, as well as the ceiling above your oven. Inspect all of your tools before use and maintain them if necessary. If you come across any tool that is broken, discard it and replace it before beginning your preservation recipe.

Be particular about tools that you use for mixing, storing, and otherwise interacting with your food, too. You might think it is only the large appliances that pose a threat, but that is not the case. A crack in a jar, a ladle that is separating from its handle, or any other cracks or breaks in small kitchen tools can harbor dangerous bacteria and contaminate your food with that bacteria. Even if you attempt to wash it properly or heat it to a safe temperature, bacteria can linger in tough cracks and become a danger to your cooking process.

Use the Freshest Foods Possible

Always use the freshest food possible, and avoid using any foods that are showing signs of spoilage or that are soon to expire. Using foods too close to their expiry date can result in them having a chance to develop harmful bacteria before you have a chance to preserve them. Once you preserve those foods, those harmful bacteria can multiply and destroy your food. In a best-case scenario, you will have an incredibly gross tasting preserve. In a worst-case scenario, you will end up with a life-threatening illness caused by eating contaminated food. Likely, however, you will end up with a lot of wasted food on your hands.

Cook Foods All the Way Through

Standard cooking rules still apply when it comes to preserving foods. With meats, in particular, you need to be confident that you have cooked them all the way through. Even if you are using less common cooking methods like smoking your meat or storing your tuna in oil, a high-quality meat thermometer should always be available when you are cooking any sort of meat or seafood product, and you should know how to use it properly. Always cook your meats all the way through to avoid having them retain harmful bacteria or parasites that can wreak havoc on anyone who may try to eat them.

Follow Exact Recipes That Match Safety Standards

Over the years, food preservation has been largely researched and perfected by professionals. Following food preservation recipes from these professionals is the best way to ensure that you are getting recipes that follow the latest safety measures for that unique type of food preservation. When you follow a recipe, always follow it exactly. Never double recipes or half recipes unless you are absolutely confident that it is safe to do so, as most recipes are made in such a way that keeps them exact and safe. Not following a recipe to its exact standards could result in you accidentally introducing harmful bacteria to your preserve, or allowing for harmful bacteria to thrive in that environment, which can lead to food spoilage and possible illness.

Properly Label Everything You Preserve

Every time you preserve something, label it. Even if you think you will remember what it is and when you preserved it, label it. Ideally, you should put a "made on" date and a "best by" date on every single thing you make. Labels ensure that you always know what is inside of a container, when it was made, and when it should be consumed by. A permanent marker and labels purchased in the canning section of any major department store should be plenty for most food preservation labeling needs. If you want to avoid cleaning stickers off of things, later on, choose labels that dissolve in water.

Use the Right Packages to Store Your Preserves

Just like you need to follow the right methods for preparing your foods, you also need to follow the correct methods for storing your foods. Avoid storing your foods in anything other than what a recipe calls for, or what you know to be absolutely safe for a certain method of preservation. Storing your food in the wrong container could expose it to excessive amounts of light, oxygen, moisture, heat, or pests. Never size up or size down on a package unless a recipe says it is safe to do so, or that particular preservation method is known for being safe to do so.

Store Your Preserves In the Right Conditions

Aside from keeping your preserves in the right containers, you also need to keep them in the right conditions. Ensure that they are kept in a proper, cool, dry place where they can be kept for long periods of time. The best space in your house to store your preserved foods is a root cellar. If you do not have a root cellar, a dark, cool closet away from the kitchen is ideal. Avoid a closet too close to your kitchen as they have a tendency to heat up from the warmth of cooking, and this can deteriorate the quality of your preserves.

Keep Track of and Consume Your Preserves In Proper Timing

After you have labeled foods and safely stored them away, you should always make a note of what you have preserved. Keeping an "inventory list" of all of the foods you have preserved is a great way for you to quickly see what you have on hand and use it before it spoils. Your inventory list should include all of the food items you have, organized by category, as well as the dates they must be consumed by. Keep those that need to be consumed quickest at the top of your list, so you know when they need to be eaten by. Plan your meals around this list and your inventory so that you use all of your preserves before they spoil. This way, you do not end up wasting any of the food you put so much effort into saving for your family.

Research a Method Before Actually Trying It

Insufficient research and improper understanding can lead to you making mistakes during the preservation process, which could result in serious illnesses or injuries. Be sure to completely read through the steps on how to complete a certain preservation method beforehand, educate yourself on the safety measures of that method, and learn the recipe before you try it. If you are extremely new to something, having the help of someone who is already experienced in the said method can be a great way for you to learn how to properly complete that method. Ensure that the person you are learning from follows the proper safety standards on that method so they can teach you those standards, too.

Safety Measures and Considerations When Canning Food

Canning is one of the most popular food preservation methods to date. With canning, the biggest risk aside from food spoilage is botulism. Improperly canned food items can carry high levels of botulism, which can infect people and cause illness and even death. Although overall death rates from foodborne botulism are low, you still need to be incredibly careful when using this preservation method to prevent the risk of infection. The lowered rate of infection is likely due to the fact that we now have a clear understanding of what safe canning looks like and how to completely eliminate botulism spores from the food contents that are stored within the can.

Always Use the Right Canning Method

Never try to can foods using a canning method other than the one described in the recipe. Attempting to water bath can recipes that should have been processed in pressure canners, or pressure can recipes that should have been processed in water bath canners, can lead to dangerous side effects. For foods improperly canned in water bath canners, the highest risk is with botulism infection. For foods improperly canned in pressure canners, the highest risk is that the pressure canner explodes and causes serious damage to your home and possibly serious injury to you or anyone around your pressure canner.

Use the Right Level of Acidity

Canning recipes call for certain levels of acidity to ensure that they are able to remain shelf-stable. The acidity of the contents in the can directly contribute to the foods' ability to remain safely preserved for lengthy periods of time. Even in low-acid foods, salts and other acidic preservatives are added to the recipe to ensure that you are storing the contents as safe as possible.

In a water bath canner, it is especially important to check your acidity levels since it is intended *only* to be used with high acidity foods. To ensure absolute safety, you should use ½ teaspoon of citric acid or two tablespoons of bottled lemon juice per quart of tomatoes. If you are canning pints, you will need one tablespoon of bottled lemon juice or ¼ teaspoon of citric acid per pint.

Use New Lids Every Time

Canning jar lids are not intended to be used more than once, unless you are using them on dry sealing methods, such as storing dehydrated foods or dry ingredients. Attempting to reuse a lid through a canning processing experience can result in it not properly sealing, which can lead to food spoilage and the introduction of botulism spores in the food you are preparing. You should always purchase new lids every time you are canning, and you should use new screw bands any time your screw bands start to rust. Otherwise, the rusted bands will be challenging to remove, and you will not be able to retrieve your contents from inside the jar. There is one long-term alternative to replacing jar seals, which is to purchase reusable jar seals. These special lids are designed to be used multiple times, though they come at a much higher price. If you are willing to foot the bill, though, they last longer and are more environmentally friendly. Be sure you use, store, and clean your reusable lids exactly to the manufacturer's recommendation to avoid accidental illness.

Always Follow the Recommended Headspace Rule

The amount of headspace recommended in a canning jar is developed to minimize the amount of air in each jar without running the risk of the jar exploding. Certain recipes, like salsa, will require larger headspace because of how much pressure they generate. Other recipes, like jam, will require less headspace because they do not produce as much pressure inside of the jar. If you leave too much headspace, the seal may not form properly, and there is too much air, which can lead to foods turning rancid inside of the jars.

Safety Measures and Considerations When Dehydrating Food

Dehydrating food is a popular method among avid food preservation fans, as well as people who love to hike. In the prepper community, this is a preferred method for storing food for as long as possible without having to follow all of the intense requirements of canning foods. As well, many simply enjoy the taste of dehydrated food.

Spread Everything Out In Shallow Layers

When you are dehydrating food, ensure to spread everything out in shallow, single layers. Piling things up or allowing the layers to become too thick can result in you having food that does not dry out fast enough, which can pose a serious threat. Food that does not dry out fast enough can start to develop harmful bacteria during the dehydrating process, which results in the food not being safe for consumption.

Store Everything In Air Tight Containers

Dehydrated food always has the ability to rehydrate by capturing moisture out of the air or anything else that may come into contact with it. Keeping your food in airtight containers is essential to avoid the food from growing dangerous bacteria. Even if you are going to be freezing your dehydrated foods, you need to keep them in airtight containers to avoid them rehydrating from the ice in the freezer. Vacuum sealed bags are the best choice for dehydrated foods that are not dehydrated to the point of being brittle.

Keep Everything In the Right Storage Conditions

Foods that are dehydrated to the point of being brittle, such as potato flakes or herbs, can safely be stored in airtight containers at room temperature. Ensure that these are kept in the smallest jar possible so that not a lot of air can get in with them or, better yet, store them in vacuum-sealed bags. If the food you have dehydrated is not brittle, such as jerky or mango slices, you will need to vacuum seal it and store it in your freezer. While these can be stored on the shelf for a period of time, room temperature will drastically reduce their lifespan, resulting in them developing harmful bacteria and turning rancid.

Rehydrated Foods Are Perishable

If you are going to be creating dehydrated foods that you plan to rehydrate later, such as potato flakes, it is important to note that the rehydrated foods are perishable. If you store your foods improperly and they become rehydrated in the process, they should be considered perishable, too. Since there is no way of knowing when the moisture was reintroduced, it would be best to toss that food out and start over.

Safety Measures and Considerations When Freezing Food

Freezing food is something most of us do, so it seems strange that there would be unique considerations for you to be mindful of when it comes to freezing foods. After all, it seems like common sense, doesn't it? There are, however, a few things you need to consider when it comes to storing food items in your freezer.

Your Freezer Must Always Be Kept At the Right Temperature

In order for your freezer to work properly, it must be kept at a temperature less than 0F. If the temperature rises about 0F, the food inside of the freezer will become dangerous to eat because it has been exposed to conditions where parasites, bacteria, and mold can begin to grow on that food. If the temperature remains below 0F the entire time, food can be stored indefinitely.

Freezing Does Not Kill Off Any Parasites, Bacteria, or Molds

Although freezing will protect your foods, it does not kill off any parasites, bacteria, or mold that have begun to grow within your food. If you freeze your food, the only thing that happens to the parasites, bacteria, and mold is that it becomes inactive. As soon as the food is defrosted, the parasites, bacteria, and mold will become active again and will continue to contaminate the meat. For certain parasites and bacteria, the cooking process will kill them off, and the food will become safe to eat. For other parasites and mold, they will continue to remain and contaminate the food regardless of what is done after they are defrosted, so the best option is to throw them away.

Food Must Be Frozen At the Right Time In Its Lifecycle

100

Food that is frozen at the peak of its lifecycle retains all of the nutritional content that it had before going into the freezer. It also continues to taste great, and works excellent in most dishes, as long as it has not become freezer burnt or been stored for too long. It is best to store food at the peak of its freshness, rather than at the end of its lifecycle, to ensure it remains tasty and healthy to eat.

Proper Packaging Must Happen to Prevent Food Deterioration

Properly packaging your food is essential if you are going to prevent the food from deteriorating. You need to prevent the frost and ice in your freezer from coming into direct contact with the food if you are going to be able to prevent freezer burn and discoloration. While freezer burn and discoloration do not render food inedible, and it will still be perfectly safe to eat, the tastemay be off. In some cases, extensive freezer burn and discoloration can lead to the food tasting terrible.

Safety Measures and Considerations When Brining and Salting Food

Brining is something that many people do as a way to tenderize their meat and improve the flavor. In the case of smoking food, brining or curing the meat is an essential step to higher quality meat that dehydrates and preserves better. Following proper safety measures is essential when brining food.

Always Use the Proper Salt Ratio

Always use the right amount of salt when you are brining. First and foremost, a brine will not work if there is not enough salt, as salt is responsible for the osmosis process that happens in the meat while it brines. Second, salt is what helps prevent your food from developing harmful bacteria. Foods that would generally go rotten in the fridge in just a couple days can be stored for a week or more in brine, depending on the recipe you are using. The way this is possible by using the proper amount of salt, always.

Ensure That All of Your Food Is Properly Covered

If any of your food is sticking above the salty brine, it is at risk of developing harmful bacteria or mold. In some recipes, they will suggest you salt the exposed top to prevent this from happening. In others, they may require you to use a weighted item to keep that food below the surface of the brine. Ensure that you follow these directions clearly so that you are not putting your food at risk of developing harmful bacteria.

Always Use Food-Grade Materials When Brining Food

Stainless steel, glass, and high-quality BPA-free, food-grade plastic are the best for brining. Ensure your bowls, crocks, and utensils are all made of these materials to avoid accidentally introducing harmful bacteria into your food.

Store Your Brines In the Fridge

In days gone by, brined meats were stored in cool root cellars where they stayed cold enough to remain preserved for extended periods of time. These days, the average house does not have a root cellar. As well, the root cellars are not as safe as the fridges we do have. Always store your brines in the fridge to avoid the salt brine and food from heating up to the point where it becomes dangerous to consume.

Let Your Meat Come to Room Temperature Before Cooking

Before you cook meat that has been preserved in a salt brine, always let it come up to room temperature. Cooking cold meat in cooking oil can result in serious spitting and, in some cases, explosions in your kitchen. Letting the meat warm up first not only helps tenderize it, but it also avoids you experiencing a dangerous explosion in your kitchen.

Safety Measures and Considerations When Sugaring Food

Sugaring is a fascinating method, yet it is rather hard to find information on in the modern age. Still, there are things you should know about safely sugaring food so that it stays preserved.

Always Use the Recommended Amount of Sugar

Ensure that you always use the recommended amount of sugar when sugaring foods, and only sugar foods that have specific recipes available. It might seem like you are using a lot of sugar, but that sugar content is essential for proper storage. Do not adjust the ratio. If the sugar seems like too much, use an alternative storage method.

Store Your Sugared Foods In the Fridge

Sugared foods are not shelf-stable. There is not enough acidic content in them for them to be properly preserved at room temperature. So you should never attempt to do so. The purpose of sugaring foods is to extend their available lifespan in the fridge.

Safety Measures and Considerations When Smoking Food

Smoking food is a rather popular concept in the modern age. It tastes great, and it is fairly easy to do when you know what you are doing. The following safety tips will ensure that you are able to preserve all of your foods through your smoker safely.

Use Proper Safety Equipment

Smokers get incredibly hot, and they produce a large amount of smoke. The smoke itself can become extremely hot and can produce burns if you are not careful. Always use the proper safety equipment and safety measures when you are working with a smoker. If you need to open the smoker to check on your foods, the appropriate method is to open the smoker and stand back while the excess smoke dissipates. Then, you can move forward and check your meats. Meats in a smoker do not need to be turned, so do not do this. Doing so only puts you at risk of burning yourself from unnecessarily tampering with the food.

Use the Proper Amount of Wood Chips

Using too many wood chips can lead to you having added fuel in your smoker, which can lead to a fire. Refrain from adding extra wood chips or chunks to extend your smoke time and instead commit to adding more as needed.

Do Not Leave the Smoker Unattended

Smokers are hot, and they produce flames to keep the wood smoke available for your food. For this reason, they are a fire hazard. Ensure that your smoker is on sturdy, non-flammable ground, and remain near your smoker at all times so that you can watch it and take immediate action if anything goes wrong. The user manual will tell you what to do in the event of a fire. Be sure you read that before using your smoker so that you are prepared in case of an accident.

Safety Measures and Considerations When Pickling and Fermenting Food

Pickling and fermenting foods are something you have to be careful with because you are relying on salt to prevent food spoilage. Improper methods could lead to botulism poisoning. The following tips will help you safely pickle and ferment foods.

Only Follow USDA-Approved Recipes

USDA-approved recipes provide adequate salt ratios, preparation guidance, and storage measures for pickling or fermenting foods. Ensure that you follow USDA-approved recipes as they will contain everything you need to know, including up-to-date safety measures. Avoid heritage recipes that may lack proper safety standards to avoid botulism or other types of potentially fatal food poisoning.

Use the Proper Amounts, and Types, of Salt

Like brining, pickling, and fermenting both require salt to inhibit the development of bacteria. Use the proper amounts and types of salt to avoid the development of unwanted bacteria.

Store Them Per the Recipe's Recommendation

Each pickling and fermenting recipe will have specific recommendations on how to store that recipe. Usually, it will include placing them in the fridge for several months or canning them for up to a few years. Follow the exact recommendations to ensure your food remains free of harmful bacteria after being processed.

Safety Measures and Considerations When Ash, Oil, and Honeying Food

Ash, oil, and honey are old-school methods for preserving food. They work excellently, but you need to ensure you have the right safety practices in place to avoid food spoilage, or worse.

Follow Exact Recipes

To avoid accidental illness, always follow exact recipes. Ensure your recipes are updated to include modern food safety practices so that you are not accidentally introducing bad bacteria into your foods. In addition to following exact recipes, follow exact storage methods, too. And, always use the freshest wood ash, oil, and honey available for long term storage.

Check on Your Food Regularly

Food preserved these ways should be fine, but as with any preserved food, there is always room for error. Check on your foods every so often to ensure they are still safe to consume. If they are not, remove and discard the poor quality foods, so they do not contaminate everything else.

Chapter 15

Juice and Smoothie Recipes

Being able to preserve your harvest is important. Still, it is also essential to know how to use your preserved harvest to help you create delicious meals. During times of survival, especially, eating extremely nutritional meals is important as it helps keep you sustained through stressful periods. The added nutrients will also help if you find yourself having to put more work into your daily life so that you can sustain your survival.

Although working with preserved foods is different from working with natural foods, there are still many simple and delicious recipes you can make that will allow you to enjoy the foods you have preserved.

One of the best foods you can make for yourself out of your preserved harvests is drinks. Smoothies and fresh juices are nutritious, healthy, and can give you a large amount of protein through a relatively small amount of food. They are simple to make, easy to digest, and can offer many benefits. Plus, if you are someone who tends to struggle with eating when you are stressed, smoothies and juices go down well, which makes it easier for you to load up on proper nutrition during difficult times.

Honey Berry and Spirulina Juice

Makes 1 serving of juice.

What You Need:

- 2 TBSP honey berry puree

- 1 TSP spirulina

- 1 cup of filtered water, milk, or fruit juice

How to Make It:

1. Blend all of your ingredients together in a blender. Taste it for sweetness; if you want it sweeter or more flavored, you will want to add more of your honey berry puree.

Strawberry Banana Oat Smoothie

Makes 1 smoothie.

What You Need:

- 1 frozen banana
- ½ cup frozen strawberries
- 2 TBSP soaked oats (soak overnight for softest texture)
- 1 TBSP peanut butter
- 1 cup filtered water or milk

How to Make It:

1. Cut your frozen banana into 1" slices and then place everything in your blender. Blend until you reach your desired consistency.

Frozen Peach Raspberry Dream Smoothie

Makes 1 smoothie.

What You Need:

- ¼ cup canned peaches
- ¼ cup frozen raspberries
- 1 TBSP lemon juice

- 1 cup filtered water or tropical juice

- 2 ice cubes

How to Make It:

1. Blend everything in your blender until you reach your desired consistency. If the taste of your smoothie is tart because of the raspberries, add some of the syrup out of your jar of peaches.

Superfruit Smoothie

This recipe makes 1 smoothie.

What You Need:

- ¼ cup vanilla yogurt

- ½ cup frozen berries (whatever mix you like)

- 1 cup ice

- 1 TSP coconut sugar

- Filtered water to reach your desired consistency

How to Make It:

1. Add everything to your blender. Start with just ½ cup of water and increase your water content from there until you reach your desired smoothie consistency.

Super Banana Oat Smoothie

Makes 1 smoothie.

What You Need:

- ¼ cup rolled oats

- ½ cup plain yogurt

- 1 frozen banana, sliced into 1" pieces

- ½ cup milk

- 2 TSP raw honey

- ¼ TSP cinnamon

How to Make It:

1. If you prefer softer oats in your recipes, combine the oats and milk and let them sit in the fridge overnight. Then, place everything in your blender and blend it until you reach your desired consistency. Adjust your honey content to reach your desired level of sweetness.

Ultra Green Smoothie

Makes 1 smoothie.

What You Need:

- 1 frozen banana, cut into 1" slices

- ½ cup blueberries, sliced

- ½ cup almond milk

- ½ cup baby spinach

- ¼ cup plain yogurt

- 1 TBSP nut butter, any flavor you like

- 3 mint leaves

How to Make It:

1. Add all of your ingredients to your blender and blend until you reach your desired consistency. Add extra milk to adjust thickness if you find that your smoothie is too thick.

Super Zinger Breakfast Juice

Makes 1 cup of juice.

What You Need:

- 2 carrots, peeled and chopped
- 2 beets, trimmed, peeled and chopped
- 2 apples, peeled, seeds removed and quartered
- 2 lemons, peeled, seeds removed and quartered

How to Make It:

1. Place everything in your juicer and let it run into a large glass. Add some raw honey to sweeten the drink if you are not happy with the flavor. Use your honey berry puree for an even sweeter and more distinct flavor if you wish.

Slam It Down Kale Juice

Makes 1 cup of juice.

What You Need:

- 1 stalk celery, chopped into 3" lengths
- 1 cucumber, chopped into 3" lengths
- 5 kale leaves
- 2 TBSP parsley

- 3-4 pieces of pineapple

How to Make It:

1. Cut your vegetables and fruit into sizes that will fit into your juicer. Run them through your juicer and let them fall into a large cup. Adjust pineapple for more or less sweetness in your juice, or add an apple for additional sweetness.

Chapter 16

Meal Recipes

Aside from smoothies and juices, proper meals are important to eat, too. You should be eating at least three wholesome meals every single day to ensure that you are getting plenty of sustenance to keep you going.

Because you have already done so much work in preserving your food, a lot of the time preparing a meal from your stockpile is as simple as opening a few jars and warming everything up. Of course, there are many great ways that you can pair these preserves together to ensure that you are getting a proper nutritional intake.

It is important that you focus on continuing to eat all of the important foods every single day, even if you are relying heavily on your stockpile. This means every day you should be eating hearty servings of fruits and vegetables, meat, and grains. Keeping a healthy array of proteins, fats, and carbohydrates on your plate for each meal will ensure that your body has everything you need to sustain yourself. Ensure that you regularly choose different types of vegetables, meats, and grains out of your pantry so that you get a varied diet. Variation in your diet is essential to helping you get all of the different types of vitamins and minerals that you need on a day to day basis.

One thing you should be cautious of when eating out of your preservation stockpile is portions. When you are not in a survival setting, it can be easy to overstock everyone's plates and eat more than you truly need. Aside from this being generally unhealthy, it can also lead to you depleting your stockpile much faster than you need to. Rather than letting your stockpile get depleted, make sure everyone gets proper rations. Give people the recommended daily amount, no more and no less. This way, everyone can sustain themselves, and your stockpile lasts as long as possible.

To help inspire you to get started with using your preservation stockpile in recipes, I have included a supply of recipes below. These will show you exactly how you can transform your preserved foods into delicious meals that will keep you and your family going for extended periods of time.

Mashed Potatoes and Meatloaf

Makes 2 servings of mashed potatoes and 1 whole meatloaf.

What You Need:

- 2/3 cup potato flakes
- 2/3 cup water
- ¼ cup milk
- 1 TBSP butter
- ¼ TSP salt
- 1 lb ground beef, defrosted
- ½ cup breadcrumbs
- ¼ cup tomato sauce
- 1 egg
- 2 TBSP dried herb mix

How to Make It:

1. Boil 2/3 cup of water. Add potato flakes until they are rehydrated. Remove the mixture from heat and add milk, butter, and salt. Let stand for 10 minutes, then whisk again for a fluffier texture.

2. Meanwhile, preheat the oven to 350F. In a bowl, mash together beef, breadcrumbs, tomato sauce, egg, and dried herbs. Using your hands, shape the beef mixture into a meatloaf and place it in a bread pan to bake—Bake the meatloaf for 30 minutes, or until it reaches safe internal temperatures.

Kielbasa and Sauerkraut

Makes 4 servings.

What You Need:

- 3 potatoes, peeled and diced
- 16 oz sauerkraut
- 1lb kielbasa sausage, cut into ½ inch lengths
- 1 onion, sliced thin
- ½ cup butter
- 2 garlic cloves, minced
- ½ TSP thyme
- ¼ TSP sage
- ¼ TSP black pepper

How to Make It:

1. Heat cold butter in a pan with the onions. Simmer over medium heat for 10 minutes. Add garlic, thyme, sage, and black pepper. Let it simmer for 2 minutes, so herbs become fragrant.

2. Add sauerkraut with its liquid contents, kielbasa pieces, and potato chunks. Simmer.

3. Pour mixture into a casserole dish and bake at 225F for 3 hours.

Crockpot Chili

Makes 16 servings.

What You Need:

- 1lb ground beef

- ½ onion, sliced

- 16 oz stewed tomatoes, with juice

- 8 oz tomato sauce

- 3 TBSP dried herbs

- 1 cup dried beans, any mix you like

- 3 cups soup stock, beef or vegetable

How to Make It:

1. Brown your ground beef with 1 TBSP of your dried herb mix. Add the cooked beef to your crockpot. If you'd like, you can sauté your onion in the retained beef juices. Otherwise, just add them fresh.

2. Combine all of your ingredients in a crockpot and cook over low heat until the beans are cooked all the way through. This takes about 3-5 hours.

3. Consume what you can in two days, freeze the rest in one or two serving increments.

Tuna Sandwiches

Makes 2 tuna sandwiches, plus extra bread.

What You Need:

- 1 jar of tuna in oil

- ¼ cup mayonnaise

- 3 cups flour

- 1.5 TBSP melted butter

- 1.5 TBSP sugar

- 1 TSP salt

- 1 TSP yeast

- 1 cup whole milk, lukewarm

- ¼ cup warm water, 115F

How to Make It:

1. Dissolve one-half tablespoon of the sugar in the warm water and add the yeast. Let it sit for 15 minutes until the yeast becomes frothy.

2. Mix 1.5 cups of flour, the remaining sugar, and salt together. Add the yeast mixture and stir until it for about 5 minutes, until no lumps are remaining.

3. Add the remaining flour and knead it for 10 minutes. Let it rest for 15 minutes. Transfer it into a floured bowl and let it rise for 2+ hours. The longer, the better your bread will taste.

4. Punch the dough down, knead it for 5 minutes. Place it in a greased loaf pan and let it rise for 2+ hours. Again, the longer the rise, the better the bread.

5. Bake the bread at 350F for 30 minutes, or until the top is golden brown. Remove it from the oven, rub the top with melted butter, and let it rest for half an hour.

6. Drain the tuna, flake it with your fork, and add mayo. Mix it together until you have a wet, sticky tuna salad texture.

7. Slice your bread and place tuna salad on one side of the bread. Top it with another piece and serve!

8. Store leftover bread in saran wrap and tinfoil in the cupboard and eat it within three days for the freshest flavor.

Hearty Cowboy Trails Dinner

Makes 2 meals.

What You Need:

- 1 cup dried kidney beans
- 1.5 + 2/3 cups water
- 1.5 tsp salt
- 2/3 cup potato flakes
- ¼ cup milk
- 1 TBSP butter
- ¼ TSP salt
- 4 strips beef jerky

How to Make It:

1. Soak your dried beans in 1.5 cups of water and salt overnight. Keep them in the fridge to prevent mold from developing on them.

2. When you are ready to eat, boil your beans for about 1.5 hours, or until done. Top the pot up with water as needed to prevent it from boiling off. (Alternatively, cook dried beans in 6 cups of water in the crockpot until they are done, about 3-5 hours.)

3. Thirty minutes before your beans are done, boil 2/3 cups of water. Add potato flakes and cook until they are rehydrated. Remove from heat, add milk, butter, and salt. Let stand for 10 minutes, then fluff it with a fork.

4. Serve ½ cup beans, ½ cup mashed potatoes, and two slices of beef jerky on each plate. Eat fresh.

Conclusion

Congratulations on reading *Survival 101: Food Storage*. I wrote this book to help you discover how you can safely and effectively preserve your food so that you can develop a hearty stockpile for your family. During uncertain times when grocery stores are sold out, and supply chains are running dry, knowing how to preserve your own food is important. Through preserving your own food, you ensure that your family has plenty to eat regardless of what is going on in the world around you.

For many, taking direct control over your food supply, as well as your survival, is a critical way to help you feel safe and confident during such challenging times. This way, no matter what is going on in the world, you know your family can rely on you to remain safe and healthy through the experience.

Preserving your own food can seem like a daunting task if you are new to it. I suggest you pick just two preservation methods to start with and begin there. Preserve as much as you can using these two methods. Ensure that you follow adequate safety measures and that you prepare as much as you possibly can for your family. If, by the time you are done, you feel ready to preserve more food using an alternative method, you can add that to your list. This way, you do not become overwhelmed, and you are able to confidently and safely preserve your food in minimal timing.

When you are preserving food, you must follow exact safety measures to ensure that your food remains safe to consume. Improperly preserved food can lead to serious illness, such as botulism, which can be fatal. The USDA has lists of certified recipes that are known for following proper, modern safety measures to ensure that you are using the best recipes possible. Understand that heritage recipes may have worked for people in the past, but those people did not know what we know now. As a result, many of them fell ill from consuming their food. These days, we have access to proper research, science, and technology that allows us to safely preserve and store foods without risking our health, or the health of our family. If you do choose to use a heritage recipe, compare it to USDA standards and adjust it as needed to ensure that your recipe is safe for use.

A great way to help yourself preserve recipes safely is to find an experienced person that can help you. Someone who can show you the ropes is a great way to have access to all of the support you need to properly and safely preserve your food. Ensure that the person you are learning from is up-to-date on the latest safety standards and that they use them in their own preservation methods, too, so that you can confidently consume everything you prepare. If you cannot find someone in your local area that can help you, consider finding someone online who can tell you everything you need to know. It may not be as hands-on, but it is still a great way to learn everything you can.

If you are interested in learning more about survival, I encourage you to check out my three other books on this very subject. *Survival 101: Raised Bed Gardening, Survival 101: Bushcraft,* and *Survival 101: Beginner's Guide* are all great books designed to help you take your survival into your own hands. This way, no matter what, you feel confident in your ability to keep your family safe and alive.

Before you go, I ask that you please take a moment to review *Survival 101: Food Storage* on Amazon Kindle. Your honest feedback would be greatly appreciated, as it helps me create more great content for you.

Thank you, and best of luck. Stay safe out there, friend!

www.ingramcontent.com/pod-product-compliance
Lightning Source LLC
Chambersburg PA
CBHW080421030426

42335CB00020B/2534